**WITHDRAWN**

# Seen *and* Unseen

LEARNING RESOURCE CENTER
NORTH PLATTE COMMUNITY COLLEGE

D1558327

# A FALCON GUIDE®

# Seen and Unseen

## Discovering the Microbes of Yellowstone

Kathy B. Sheehan

David J. Patterson

Brett Leigh Dicks

Joan M. Henson

FALCON®

GUILFORD, CONNECTICUT
HELENA, MONTANA
AN IMPRINT OF THE GLOBE PEQUOT PRESS

To buy books in quantity for corporate use
or incentives, call **(800) 962–0973, ext. 4551**,
or e-mail **premiums@GlobePequot.com**.

# A FALCON GUIDE®

Copyright © 2005 by Montana State University

All rights reserved. No part of this book may be reproduced or transmitted in any form by any means, electronic or mechanical, including photocopying and recording, or by any information storage and retrieval system, except as may be expressly permitted by the 1976 Copyright Act or by the publisher. Requests for permission should be made in writing to The Globe Pequot Press, P.O. Box 480, Guilford, Connecticut 06437.

Falcon and FalconGuide are registered trademarks of The Globe Pequot Press.

Text design by Casey Shain
Maps produced by the Spatial Analysis Center, Yellowstone National Park, August 2004

Library of Congress Cataloging-in-Publication Data
Seen and unseen : discovering the microbes of Yellowstone / Kathy B. Sheehan . . . [et al.].
    p. cm. — (A FalconGuide)
  Includes bibliographical references.
  ISBN 0-7627-3093-5
 1. Thermophilic microorganisms—Idaho. 2. Thermophilic microorganisms—Montana. 3. Thermophilic microorganisms—Wyoming. 4. Microbial ecology. 5. Yellowstone National Park. I. Sheehan, Kathy B. II. Falcon guide.
    QR107.S44 2004
    579.31758—dc22                                                                 2004047158

Manufactured in China
First Edition/First Printing

# Contents

Acknowledgments ... ix
Introduction ... xi
   Discoveries ... xiii
   Redefining the Limits of Life ... xiii
   What Is a Microbe? ... xv
   How to Use This Book ... xviii

## The Habitats ... 1

   La Duke Spring ... 7
   Mammoth Hot Springs ... 9
   Mammoth Hot Springs to Beaver Lake Area ... 15
   Amphitheater Springs Area ... 21
   Roaring Mountain to Norris Geyser Basin Area ... 25
   Norris Geyser Basin ... 29
   Lower Geyser Basin ... 33
   Midway Geyser Basin ... 37
   Upper Geyser Basin ... 41
   West Thumb Geyser Basin ... 45
   Mud Volcano Area ... 51

## The Microbes ... 55

   Viruses ... 59
   Archaea ... 60
   Bacteria ... 60
   Eukarya ... 68

## The Relationships ... 83

   Symbioses ... 87
   Microbial Mats ... 91
   Gradients ... 95
   Biomineralization ... 97
   Science and Yellowstone's Microbes ... 99

Glossary ... 103
Resources ... 107
   Books ... 107
   Useful Web Sites ... 107
Contributors ... 108

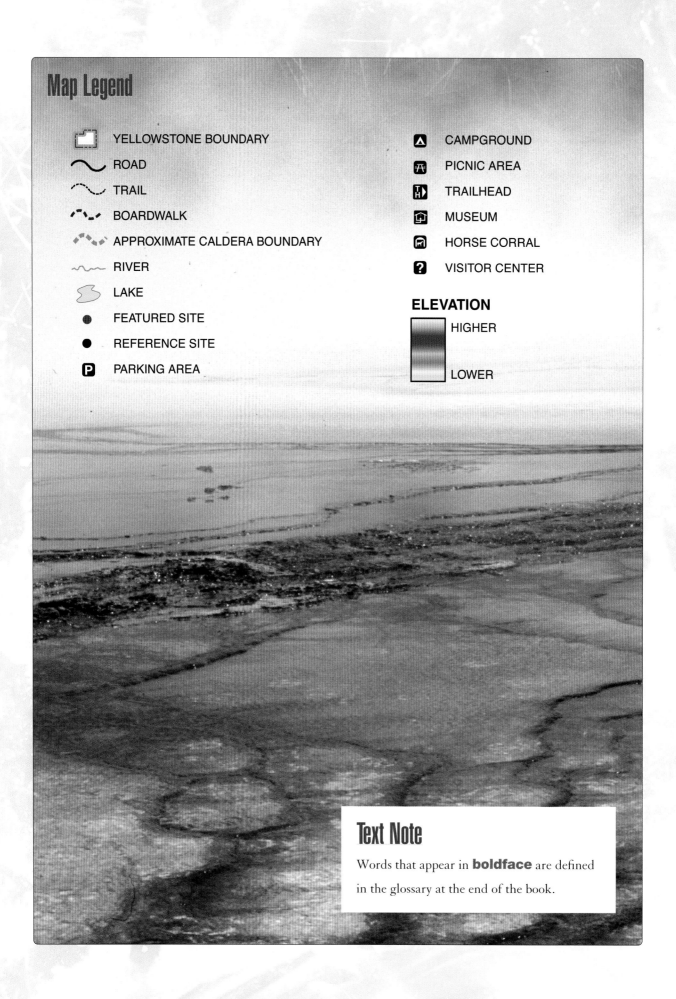

# Map Legend

- YELLOWSTONE BOUNDARY
- ROAD
- TRAIL
- BOARDWALK
- APPROXIMATE CALDERA BOUNDARY
- RIVER
- LAKE
- FEATURED SITE
- REFERENCE SITE
- PARKING AREA
- CAMPGROUND
- PICNIC AREA
- TRAILHEAD
- MUSEUM
- HORSE CORRAL
- VISITOR CENTER

**ELEVATION**
HIGHER
LOWER

## Text Note

Words that appear in **boldface** are defined in the glossary at the end of the book.

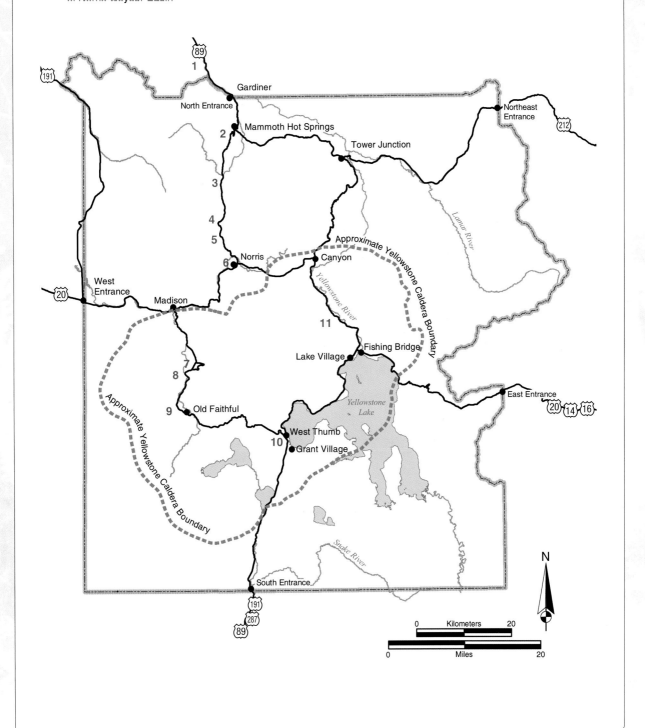

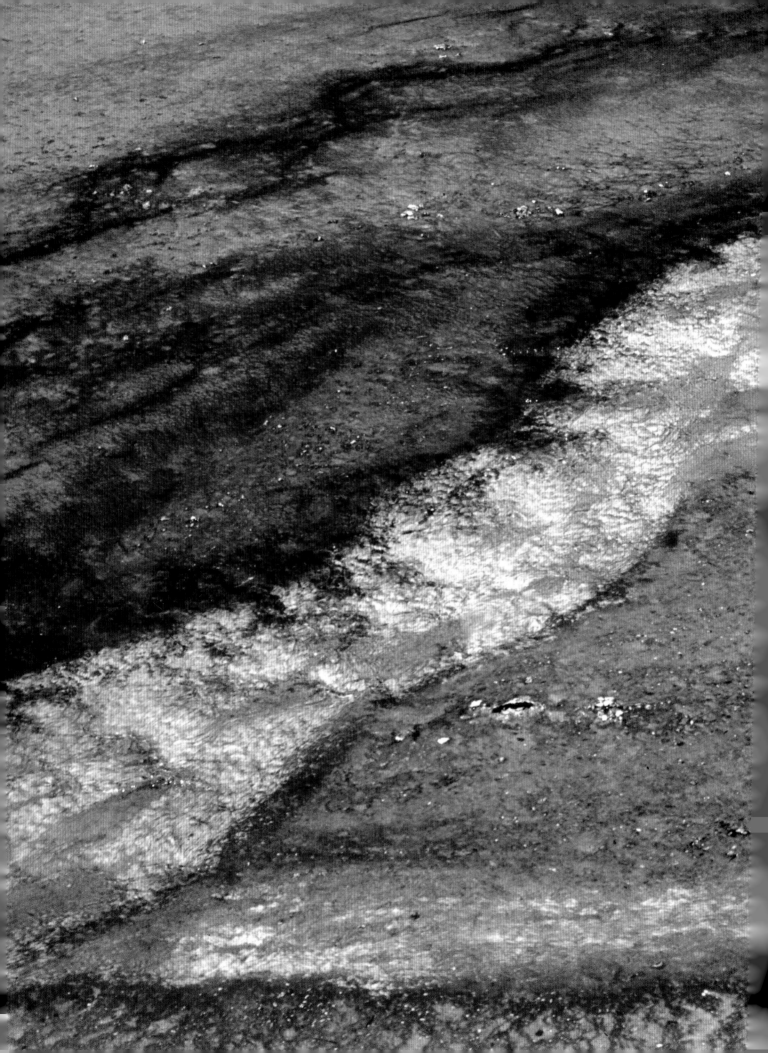

# Acknowledgments

This book was made possible with the financial support of the Thermal Biology Institute, Montana State University; the National Park Service; the National Science Foundation Microbial Observatory Grant #9977922; and the NASA Astrobiology Institute node at the Marine Biology Laboratory, Woods Hole, Massachusetts.

We are very grateful to Carl Zeiss MicroImaging, Inc., Thornwood, New York, for providing microscope equipment and an Axiocam digital camera.

Special thanks to Roger Anderson, Yellowstone National Park, for his guidance, expert editing, patience, and unwavering enthusiasm for this project. We are very grateful.

Thanks to Santa Barbara City College and the University of Sydney.

Thanks to John Varley, Christie Hendrix, Elizabeth Cleveland, Tami Blackford, Sarah Stevenson and the staff of the Yellowstone National Park Center for Resources for their valuable assistance and Shannon Savage and Adam Kiel in the Spatial Analysis Center for preparing the maps.

Woods Hole Marine Biology

University of Sydney

Many thanks to Rich Stout, Mike Ferris, Bill Inskeep, Mark Young, Sue Brumfield, K. Michael Foos, Emily Kuhn, Jessie Donohoe-Christiansen, Tim McDermott, Tai Takenaka, George Rice, Sharon Eversman, Rebecca Bunn, Cathy Zabinski, Jim Peaco, and Jane Jessell for generously providing assistance and/or additional photographs.

We appreciate the efforts of Carol Shively, Katy Duffy, Cheryl Jaworowski, and Henry Heasler, Yellowstone National Park, for their critical review of the manuscript.

Thanks to David Adams Thompson, Inc./Jason King for graphic design of the tree of life chart.

We also acknowledge Turner Enterprises for providing the samples from a bison rumen.

And many thanks to Mark Sheehan for believing in the book from start to finish.

Microbial samples for this book were collected following the guidelines for scientific research in Yellowstone National Park.

# Introduction

"Nothing ever conceived by human art could equal the peculiar vividness and delicacy of coloring of these remarkable prismatic springs. . . ."

—FERDINAND V. HAYDEN, GEOLOGICAL SURVEY TO THE YELLOWSTONE REGION, 1871

Yellowstone National Park is one of the world's most extraordinary places. Within its boundaries and in the country that surrounds the park are pristine wilderness areas and spectacular mountain scenery. The park teems with wildlife, offering visitors rare opportunities to observe wolves, grizzly bears, bison, and elk in their natural habitats. Vast forests and hundreds of waterfalls, mountain streams, and lakes fill its landscapes. Wildflowers burst into bloom during the short but precious summer months when snow melts from the peaks and in the valleys that form the Yellowstone plateau. During dry summers, raging forest fires demonstrate the power of nature and the fragility of the park's beauty. But it is Yellowstone's amazing geysers and hot springs—what scientists call **thermal features**—that are the biggest draw for more than three million visitors each year. There is no other place like this on Earth.

One of the most famous landscapes in the park is the Lower Falls of the Yellowstone River, which plunges 308 feet into the Grand Canyon of the Yellowstone.

Old Faithful erupts on average about every ninety-two minutes. The spectacular 130-foot-tall geyser expels thousands of gallons of boiling water in a one-and-a-half- to five-minute burst, much to the delight of the millions of visitors who view its performances. *Image provided by K. Sheehan.*

The nearly ten thousand thermal features in the park are the result of Yellowstone's tumultuous geologic past and present. Visitors often are surprised to learn that the central area of Yellowstone National Park is considered one of the world's largest active volcanoes.

The volcano is the result of the activity of a hot spot, one of about thirty in the world, which releases heat from deep within the Earth. The Yellowstone volcano has a history during the past 2.1 million years of some of the most powerful eruptions ever on Earth.

The most recent cataclysmic eruption in Yellowstone occurred about 640,000 years ago, ejected enormous quantities of **magma** (molten rock), and created a huge **caldera** (a giant crater), 35 miles wide by 45 miles long. Since the eruption, there have been thirty documented lava flows from the caldera that have created major features such as the Pitchstone Plateau. Glaciers during the last ice age also dramatically sculpted landscapes that visitors see today.

Currently the volcano is relatively calm, but the geologic record suggests that more explosive eruptions can occur. Dynamic geological forces are at work underneath the park. More than two thousand earthquakes occur each year. Heat is released continually from the underlying magma. Water from precipitation and snowmelt seeps into the ground and comes in contact with hot rock adjacent to the magma just a few kilometers below the surface, creating a landscape of steaming hot springs, bubbling mud, hissing **fumaroles** (steam vents), and the largest concentration of geysers in the world.

People who explore Yellowstone's geyser basins are often fascinated by the bright colors found in and around the thermal features and their runoff channels. These colors contribute much to the spectacle of Yellowstone. Some of the colors come from mineral deposits or from sunlight reflected from pools rich with dissolved or suspended chemicals, but many are due to pigments found in the **microbes** (extremely tiny organisms) that thrive in these remarkable habitats. The microbes, each too small to be seen without a microscope, form brightly colored **mats,** large communities of microbes growing together in layers of organic "slime." Some microbes intertwine to form long, hairlike filaments called **streamers.** The heated water, dissolved gases, extreme alkalinity or acidity, and/or harsh chemicals found in the hot springs prevent the growth of most other organisms that in less extreme environments would compete with and keep the microbes from proliferating. The microbes that are able to live in hot springs with temperatures between 45°C and 80°C (113°F and 176°F) are called **thermophiles** or thermophilic microbes. They are adapted to these environments and not only survive

Roaring Mountain provides a hint of the volcanic forces that lurk below ground in the hot rocks and crevices. Numerous fumaroles release steam and gases that kill most of the surrounding plants and create desolate landscapes.

but flourish, sometimes creating extraordinary mats and bizarre-looking formations.

## Discoveries

Many people, fascinated by the region's unparalleled scenic wonders, unique geology, and rich biology, have conducted explorations throughout the history of Yellowstone National Park. The wilderness protected within the park's borders provides scientists with opportunities to observe and study the natural environment in an outdoor "laboratory" where basic discoveries can be made.

During the late 1960s, Thomas Brock, a microbiologist, isolated a thermophilic bacterium, *Thermus aquaticus,* from the clear, scalding water of Mushroom Pool. Before Brock's search for microbes in Yellowstone's hot springs, it was assumed that no life could exist under such extreme conditions. Brock's bacterium was deposited in a laboratory culture collection and stored until the late 1980s, when biochemist Kary Mullis developed a powerful method of copying DNA called the **polymerase chain reaction (PCR).** The PCR method requires enzymes that can withstand repeated heating cycles. As it turned out, *Thermus aquaticus,* a bacterium adapted to life in a hot spring, had such a heat-stable enzyme. The subsequent isolation of the enzyme, named *Taq* (after *Thermus aquaticus*) polymerase, and its mass production made PCR possible and revolutionized biological research.

PCR is now the most widely used method for studying DNA in molecular biology, biotechnology, medical research, and law enforcement laboratories worldwide. Revenues from this industry are estimated to be more than two hundred million dollars each year. Mullis was awarded the Nobel Prize in 1993 for his efforts. Scientists continue to examine Yellowstone's hot springs for other novel microbes that may contain active enzymes or undergo biological processes that could benefit society.

## Redefining the Limits of Life

Hot springs in Yellowstone with temperatures at or near the boiling point harbor many species of thermophiles and **hyperthermophiles** (microbes that grow above 80°C, or 176°F). These organisms grow in some of the harshest environments on Earth and have challenged basic ideas of what is required for life.

Boardwalks allow visitors to safely observe the marvels of Yellowstone's thermal features without disturbing the delicate microbial communities.

This has led to intense interest by researchers who tackle fundamental questions in biology. How did life evolve? What are the limits for life? How do hyperthermophiles deal with physiological conditions that would kill most organisms? What roles do microbes play in extreme environments? Can they provide insights into the possibilities for extraterrestrial life?

Microbes are not restricted to the thermal areas of Yellowstone. They are everywhere, colonizing oceans, streams, lakes, soils, plants, and animals, where they maintain and support life. In fact, microbes are crucial; plants and animals could not exist without them. Although microbes are tiny, they are not insignificant. They make up the major portion of the **biomass** (weight of living material) on the planet. They are responsible for many of the biological processes that life depends upon, including generating much of the oxygen in the Earth's atmosphere.

Despite their importance, only a small percentage of microbes have been identified by biologists. This is due in part to problems with growing and studying microbes in a laboratory. Microbes have nutritional needs and growth requirements that sometimes are hard to identify and provide in a test tube. Many easily grown microbes are not representative of the diverse communities of organisms found in nature. Furthermore, many microbes look similar under a microscope, often making it difficult or impossible to distinguish them.

In the 1980s, Norman Pace and researchers in his laboratory at Indiana University conducted

*Top* This boiling hot spring at Norris Geyser Basin is home to diverse microbes adapted to life in extreme environments.

*Above* A thermal area near Old Faithful harbors a spectacular community of microbes.

pioneering experiments in Yellowstone. Instead of attempting to identify bacteria by growing cells or examining microorganisms under a microscope, the scientists collected water samples from Octopus Spring and used PCR to look for traces of the genetic material present in all living organisms. The material in different organisms is unique. By comparing the genetic material discovered in Octopus Spring (see page 35) with similar material from other organisms, the scientists were able to determine the kinds of organisms living in the hot spring. They found an incredibly rich microbial world growing in hot springs. The results of the experiments were significant for biologists and radically changed the way scientists at that time thought about the diversity of microbial life. Biologists had devised a powerful way to search for and identify microorganisms living in any habitat from deep in the oceans to rain-forest soils.

Microbial diversity is not limited to extreme environments. It flourishes in all places where life exists on Earth. To understand the ecology and activities of microbes, scientists continue to investigate many environments, including Yellowstone's hot springs, deep-sea hydrothermal vents, Antarctic ice, seawater, soils, and even the human digestive tract.

Yellowstone's wonders are more than meets the eye. An unexplored frontier awaits the curious in a drop of water on a microscope slide; in a sampling tube; or in the colorful mats, streamer communities, and boiling hot springs in the park. The world of microbes may provide answers to basic questions about bacterial **metabolism** (the chemical changes in a cell that result in energy production and growth) or lead to the development of lifesaving cancer drugs or help to unravel the mysteries of life. In any case, microbes will continue to captivate scientists and visitors to Yellowstone with their remarkable beauty, biodiversity, and amazing abilities to thrive everywhere.

# What Is a Microbe?

Microbes are tiny organisms. Some are less than 1 micron (one-thousandth of a millimeter) in size, or about fifty times smaller than the diameter of a human hair. Despite their small size, they are incredibly abundant and diverse. Millions can be present in a teaspoonful of pond water. Innumerable species of microbes inhabit the guts of animals from termites to humans, where they aid in digestion of food and protect against infection. They are found in all ecosystems on Earth, from frozen ice to boiling springs in Yellowstone to the tip of the human tongue. The microbial world includes the **viruses** and three major categories (or domains) of organisms: the **Archaea** (sometimes called *Archaebacteria*), **Bacteria** (sometimes called *Eubacteria*), and **Eukarya** (*Eukaryotes*). The Archaea and Bacteria collectively are referred to as **Prokaryotes.**

## ● Viruses

Viruses are simple, consisting of little more than a protein shell surrounding a piece of genetic material, either DNA or RNA. Most experts do not regard viruses as living cells because they are unable to reproduce except by infecting host cells and taking over the cell's metabolism. Viruses occur in natural habitats such as the oceans, hot springs, and soils. They have considerable impacts on communities of bacteria, preventing bacterial overgrowth and keeping ecosystems in balance. Other viruses cause diseases in humans, such as the HIV virus that leads to AIDS or the rhinoviruses that result in the common cold. Viruses are so small that they can be observed only with a high-powered electron microscope using a beam of electrons instead of a light beam to make an image. An electron microscope can resolve things about one thousand times smaller than the smallest objects that can be distinguished by a light microscope.

## Archaea

Archaea were classified as a separate group of organisms by Carl Woese and his colleagues at the University of Illinois in the late 1970s. Although archaea look like bacteria, their genetic make-up is so different from bacteria, plants, and animals that Woese devised the three-part classification for life—the Bacteria, the Eukarya, and the Archaea. Yellowstone's Obsidian Pool and Octopus Spring were among the first places that scientists found these amazing microbes growing in the boiling springs.

Archaea live in some of the most extreme environments on the planet, not just the hot springs in Yellowstone. Some live in the hot water found in deep-sea ocean vents and can grow at temperatures as high as 113°C (235.4°F). Others, called **halophiles,** live in very salty water. Certain archaea, called **methanogens,** live without oxygen in the intestinal tracts of animals such as humans, cows, termites, and marine life or in petroleum deposits deep within the Earth, where they produce methane gas as a waste product of metabolism.

Archaea do live under less extreme conditions, but they often are outcompeted by the large communities of bacteria and eukaryotes present in those environments.

Some scientists believe that microbes in the domain Archaea are similar to the first organisms that inhabited Earth billions of years ago and that the Eukarya (the domain that includes plants and animals) may have diverged from the archaeal branch of life around 1.7 billion years ago.

## Bacteria

Bacteria are microbes that are enormously plentiful in soils and aquatic environments or in association with plants and animals. This is due in part to their long history of life on the planet, their small size, and their ability to exploit many chemical compounds and sunlight as energy sources for growth. Many types of bacteria are able to live without oxygen or at high temperatures.

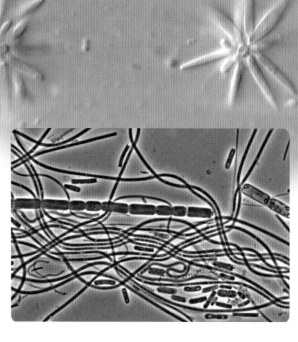

*Top* Bacteria live almost everywhere. These two colonies of bacteria were collected from the intestine of a bison, where they help to break down grass and provide essential nutrients for the bison.

*Above* These bacteria collected from a microbial mat illustrate the different shapes of bacteria, such as rods, stringy filaments, and strands formed by cells adhering end to end.

Scientists are rapidly discovering more types of bacteria by examining genetic material from many environments worldwide, including Yellowstone's hot springs. Although some species of bacteria do cause diseases or spoil food, frequently with devastating results, many bacteria are beneficial and vital for maintaining healthy balances in living systems.

Bacteria grow in a variety of shapes, from rods to stars to spiral corkscrews. They vary in size from 0.2 micron in diameter to the largest known bacterium, *Thiomargarita namibiensis*. At 750 microns in diameter, this microbe that lives in mud on the seafloor is visible with the naked eye.

Many bacteria are able to move either by using **flagella,** stiff hairlike appendages on the cell, to swim through liquids, or by gliding along solid surfaces. Some bacteria make **endospores,** dormant structures inside bacterial cells that are highly resistant to heat, dryness, radiation, and disinfectants. Endospores can survive in hostile environments for thousands, perhaps millions, of years.

## Eukarya

The major group of organisms called Eukarya includes tens of thousands of species of microscopic, single-celled organisms such as algae, protozoa, fungi, and slime molds. Eukaryotic organisms (Eukaryotes) may be multicellular too. The higher plants and animals, including humans, share similar cellular and molecular characteristics and also are classified within this major group of organisms. Plants and animals have specialized tissues, such as muscles and nerves in animals or leaves and roots in plants, and cannot survive in nature as individual cells.

Eukaryotic cells, unlike archaea and bacteria, have a distinct special compartment or **organelle,** called the nucleus. This compartment is enclosed by a membrane and contains most of their genetic material (deoxyribonucleic acid, or DNA). Cells may contain other specialized membrane-bound organelles, such as **chloroplasts** (compartments where sunlight is converted into chemical energy that can be used by the cell) that are found in algae and higher plants, or **mitochondria** (compartments where certain energy-rich molecules are produced). Eukaryotic cells usually are bigger than archaea and bacteria, ranging from 2 microns in size to 1 millimeter or more.

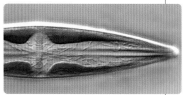
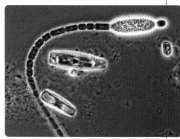

Eukaryotes are usually larger than bacteria. The microbe in the lower right-hand corner is an amoeba. Diatoms (microscopic algae) of the genus *Achnanthes* are also in the image (upper left).

*Top* Two kinds of specialized membrane-bound organelles, the greenish brown chloroplast and the stringy-looking mitochondria, are clearly visible inside this diatom, a type of microscopic algae.

*Above* These photosynthetic microbes include the long curved bacterium *Cylindrospermum*, two diatoms with brown chloroplasts, and a smaller green alga.

As a group, the eukaryotes are less tolerant of extreme environmental conditions than are prokaryotes (archaea and bacteria). Nevertheless, some microscopic algae, fungi, and protozoa are able to grow at temperatures as high as 60°C (140°F), a temperature that humans would find scalding.

Plants, algae, and some bacteria can trap sunlight energy and convert it to stored energy or "food" through an important biological process called **photosynthesis.** During photosynthesis carbon dioxide combines with water to form carbohydrates (energy-storing compounds used for food) and oxygen as a by-product. Photosynthesis requires energy-capturing substances such as **chlorophyll.** In plants, chlorophyll is located only in structures in cells called chloroplasts. In bacteria that carry out photosynthesis, chlorophyll is found throughout the cell. Chlorophyll absorbs blue and red light energy from sunlight, but not the yellow and green wavelengths. These colors are reflected from cells, giving them a green color.

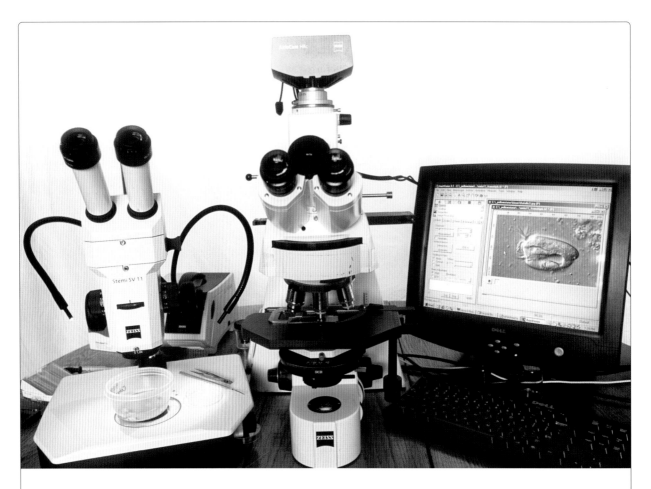

# How to Use This Book

This book describes some of the diverse microbial life found in Yellowstone. It is divided into three chapters.

The first chapter, "The Habitats," illustrates some of the accessible locations in the park where visitors can view different types of microbial communities. The second chapter, "The Microbes," examines the microbes themselves. Then, the final chapter, "The Relationships," explains some of the key roles microbes play in the communities where they are found.

Shown above are the state-of-the-art microscopes and digital camera used to collect most of the images of microbes in this book. On the far left is a stereomicroscope for examining pieces of microbial mats and plant parts. The compound microscope in the middle of the photograph has an AxioCam digital camera mounted on top and is used to view tiny samples on microscope slides. Images from the camera are viewed and stored on the computer on the right.

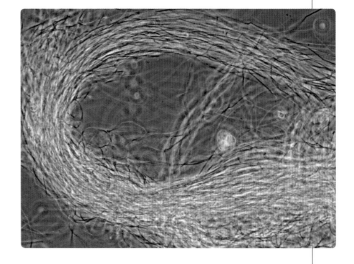

This is a highly magnified picture taken with a digital camera.

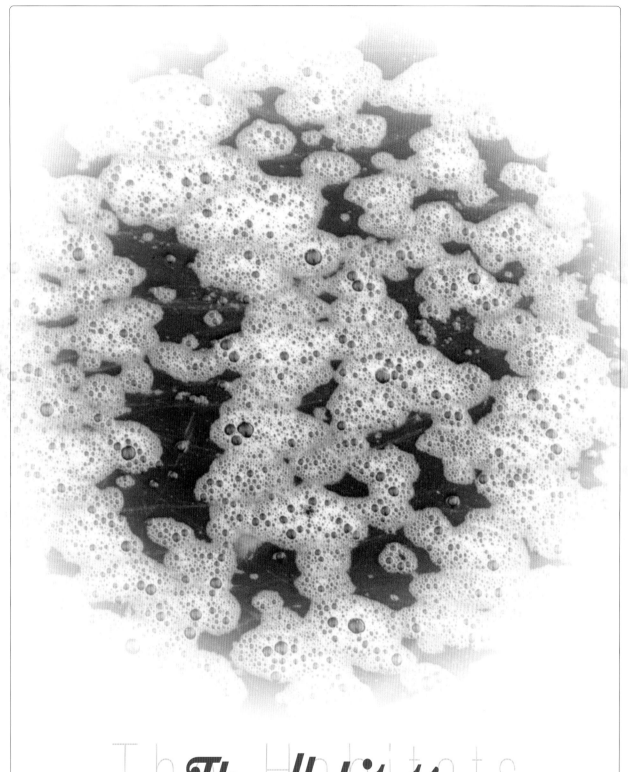

# The Habitats

Yellowstone National Park's 2.2 million acres and the areas that surround the park is one of the world's largest remaining, undisturbed, temperate zone ecosystems. The world's first national park, established in 1872, is also designated a World Heritage Site and an International Biosphere Reserve. It is home to innumerable organisms that interact with each other and the environment in self-supporting communities. The places where communities live are called **habitats.** For lodgepole pine, herds of bison, or grizzly bears, habitats extend over large tracts of land. For microbial communities, habitats can be quite small. A microbial habitat the size of a pencil eraser may contain hundreds of species each with its own requirements for life.

In Yellowstone, ecosystems are closely linked to geology. Volcanoes and earthquakes have shaped landscapes dramatically. Mountain ranges, ridges created by lava flows, plateaus, and watersheds influence everything from weather patterns to soil types, and these in turn affect plant and animal populations.

Geology impacts microbial communities too. The magma, in contact with rocks underneath the geyser basins and thermal features in Yellowstone, heats the subsurface water, as well as rain and snowmelt that seep in from the surface, to very high temperatures. This heated water dissolves silica-rich volcanic rocks. When the water reaches the surface (sometimes under pressure in an explosive geyser), it cools and silica **precipitates,** or falls out of solution from the water, as **sinter,** a grayish white material, forming cones, terraces, and rims around hot springs. In regions where the water passes through limestone, the hot water may also be enriched with calcium salts. Communities of microbes flourish in these alkaline or neutral pH thermal features, deriving energy from the mineral compounds in the thermal waters.

Other hot springs in Yellowstone contain sulfuric acid created by the metabolism of sulfur compounds by microbes or from chemical reactions when the compounds rise from deep underground to the surface and combine with oxygen. Hot, sulfur-rich environments are called **solfataras,** and they occur in other places throughout the world where there is volcanic activity. Solfataras have a characteristic smell of rotten eggs from large amounts of hydrogen sulfide gas produced by the chemical reactions. Not only are some of the springs found in solfataras acidic enough (pH below 1) to dissolve the soles of shoes, they are typically hot as well. Microbes adapted to these habitats are called **thermoacidophiles.**

Brightly colored mineral deposits occur in many of the thermal areas in Yellowstone.

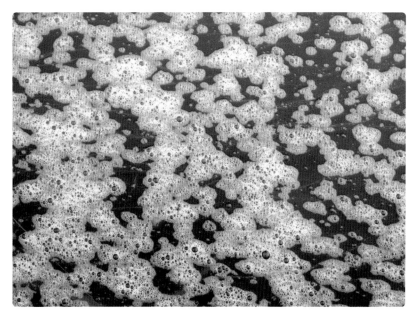

White froth, interlaced with stringy brown bacteria, grows on the surface of the water in Obsidian Creek. Organic material that blows into the creek is trapped in the froth and provides nutrients for bacteria.

Hot water creates spectacular travertine terraces as it flows down a steep hillside at Mammoth Hot Springs.

## What Is pH?

pH is a measurement of the acidity or alkalinity of a liquid. pH values range from 0 (the most acidic) to 14 (the most alkaline). Most biological processes occur in the neutral range around pH 7. Even though some microbes are able to live in environments with extreme pHs, the biological processes inside their cells still occur at near neutral pH. Even a slight change in pH inside cells can be extremely harmful.

| pH | Example |
|----|---------|
| **Acid** | |
| 0 | Concentrated acid |
| 1 | Stomach acid |
| 2 | Lemon juice |
| 3 | Vinegar, cola drinks |
| 4 | Tomatoes |
| 5 | Coffee |
| 6 | Milk |
| **Neutral** | |
| 7 | Pure water, blood |
| **Alkaline** | |
| 8 | Seawater |
| 9 | Baking soda |
| 10 | Soap |
| 11 | Household ammonia |
| 12 | Chlorine bleach, drain cleaner |
| 13 | Household lye |
| 14 | Oven cleaner |

Frequently, deep red and orange iron compounds (rust) are suspended in the springs or precipitated when steam evaporates along their edges and runoff channels. Sometimes pure sulfur precipitates, forming delicate, yellow crystals. Other minerals form hard crusts that line the basins or edges of the hot springs. Red arsenic compounds can be found in a few places like Norris Geyser Basin. At Mammoth Terraces, hot water dissolves the underlying limestone, forming white calcium carbonate–rich rock called **travertine.**

Minerals often influence the color of the light reflected from the water of hot springs. Some hot springs absorb all the colors from sunlight except intense blue light. Hot springs that contain yellow sulfur deposits often reflect brilliant green light.

Many microbes use mineral compounds as energy sources for growth. Some bacteria consume sulfur compounds and consequently contain sulfur as a by-product inside their cells. Other microbes deposit

sulfur *outside* their cells. Biogeochemical processes like these can actually aid in the formation of mineral deposits.

Each thermal feature in Yellowstone is unique with its own temperature, pH, nutrients, dissolved minerals, and communities of microbes. Each thermal feature is dynamic and changing, as well. Geological processes such as **mineralization** (the formation of hard materials such as limestone or sinter), earthquakes, and volcanic activity are constantly at work and influence everything from the frequency of geyser eruptions to the formation of terraces.

Yellowstone's thermal features are hot and potentially deadly, with boiling water and thin crusts. For safety, you must stay on the established trails and boardwalks. Avoid hot steam and spray. Never throw objects into a geyser or hot spring or disturb the microbial mats or mineral formations. Collection of any specimens, including microbes, plants, animals, and minerals, in Yellowstone is illegal and strictly prohibited without obtaining official scientific research and collecting permits and following park regulations.

The following descriptions give examples of microbial habitats in Yellowstone and are arranged along the Grand Loop Road from the north entrance of the park at Gardiner, Montana, to the Mud Volcano area just north of Yellowstone Lake.

*Top* Geological and biological processes create travertine terraces at Mammoth Hot Springs. Communities of brown and green microbes colonize the hot water flowing over the terraces.

*Below* Many hot springs in Yellowstone, like Rainbow Pool in Black Sand Basin, produce spectacular bands of color in their runoff channels.

*Top inset* Signs warn visitors that the thermal areas in Yellowstone are dangerous. People who fail to heed the warnings can be seriously injured or killed.

# La Duke Spring

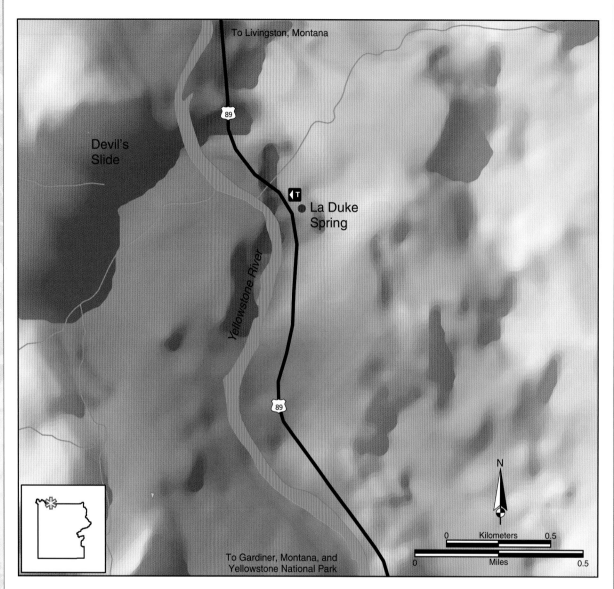

Steam rises from hot water flowing from La Duke Spring into the Yellowstone River just north of Gardiner, Montana. *Image provided by D. Patterson.*

# La Duke Spring

La Duke Spring is located on National Forest Service land about 5 miles north of Gardiner, Montana, the north entrance to Yellowstone National Park, on US Highway 89. Hot water seeps from various sources, flows along a ditch beside the highway, and travels into an old concrete holding tank. The steamy water (65°C–70°C; 149°F–158°F) is diverted through a culvert underneath the highway and down a steep riverbank, where it mixes with the cooler water of the Yellowstone River.

Layers of **cyanobacteria** (a type of photosynthetic bacterium) form colorful green and orange gelatinous mats that cover the concrete culvert and rocks. Some of the bacteria-rich water drips over the steep edges of rocks, forming stringy iciclelike structures.

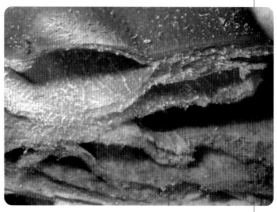
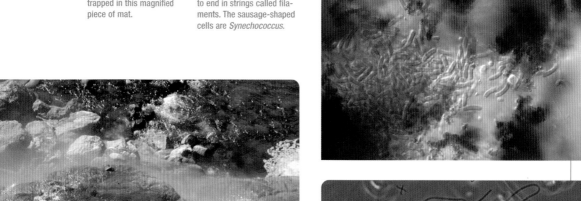

*Above* Thermophilic (heat-loving) bacteria thrive in the hot springs at temperatures that kill most organisms.

*Below* Millions of microscopic bacteria, normally too tiny to be seen, grow together in thick green and orange colonies.

*Right column, top to bottom*

The bacteria grow in slimy, jellylike layers that scientists call mats.

The microbes in the mats contain brightly colored pigments used in photosynthesis. The red flecks are pieces of iron-rich grit that are trapped in this magnified piece of mat.

*Synechococcus*, a cyanobacterium, is bright green due to chlorophyll and other pigments that capture energy from sunlight to make food by a process called photosynthesis.

This piece of mat contains the bacterium *Chloroflexus*, whose cells join together end to end in strings called filaments. The sausage-shaped cells are *Synechococcus*.

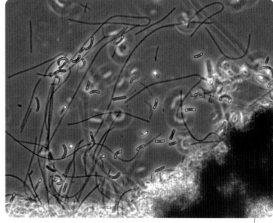

**Seen *and* Unseen**

# Mammoth Hot Springs

Impressive mineral terraces characterize Mammoth Hot Springs. These are formed when geothermally heated water flowing through and dissolving subsurface sedimentary rocks, such as limestone and dolomite, reaches the surface and degasses, and the minerals fall out of solution as calcium carbonate, a major component of limestone, creating beautiful travertine formations. The terraces are home to rich and colorful communities of thermophilic microbes. The formations are dynamic, sometimes changing rapidly when the springs are plugged up by mineral deposits. Often when a spring stops flowing, the thermophilic microbial communities die, and the color of the terrace changes to gray-white.

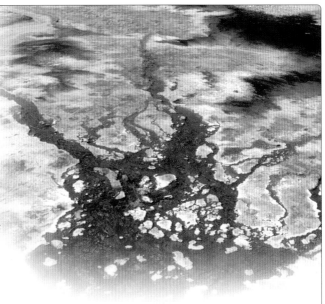

*Above* Intricate brown and green patterns come from the bacteria and algae that colonize the heated water in the hot springs.

*Below* At Canary Spring, hot water flows down a portion of steep hillside and is colonized by bright green or orange microbes.

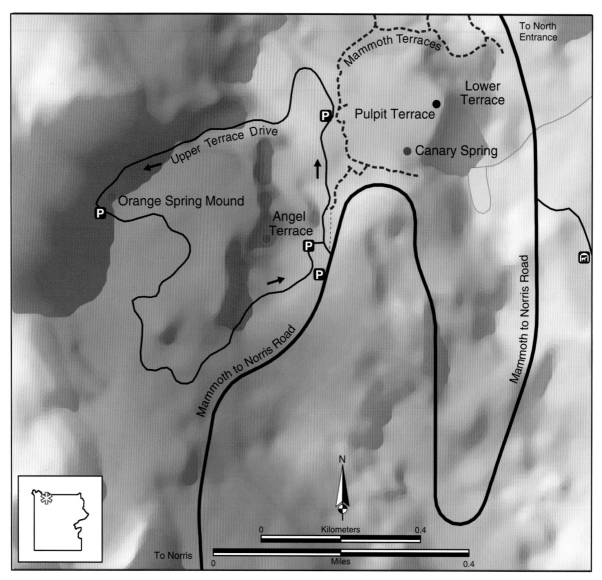

## Canary Spring

Canary Spring can be viewed from a boardwalk and viewing platforms accessed on Upper Terrace Drive or from the roadside along the Mammoth to Norris Road at the base of the terraces near the horse corrals. Water from the spring flows rapidly down a steep hillside. As the water flows away from the spring, it cools, and microbes, adapted to specific temperatures, form mats. The microbes contain bright photosynthetic **pigments** (colored chemical compounds). Microbes that live at higher temperatures have differently colored pigments than those that live at cooler temperatures. This creates patterns of colors on the hillside. Sometimes the microbes assemble into long hairlike filaments called streamers. There are excellent examples of streamers in the runoff channels of Canary Spring.

## Orange Spring Mound

Orange Spring Mound is located on Upper Terrace Drive. Springs flowing from the top of the mound produce a dramatic travertine formation draped with brightly colored communities of microbes. Green and brown layers of **diatoms** (microscopic single-celled algae) grow at the base of the mound.

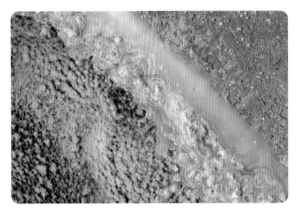

*Left* The impressive Orange Spring Mound is a large travertine formation with communities of brown microbes along its side and base.

*Top* The hot, mineral-rich water cools as it flows away from its source, providing habitats for thermophilic microbes. Minerals precipitate to form slate-blue deposits on the surface of the water.

*Center* This sample was collected at the edge of a newly flowing spring. Recently formed travertine is on the right and older travertine deposits that are colonized by green microbes are on the left.

*Above* A magnified piece of travertine.

Tiny mineral mounds, only a few millimeters across, form at the base of Orange Spring Mound around bubbles of carbon dioxide gas. Microbes quickly colonize these areas and eventually green and orange mats form. Rod-shaped bacteria, **filamentous** (threadlike) bacteria, and diatoms intertwine to form a matrix. The mat also harbors microscopic predators such as protozoa that eat bacteria and diatoms.

*Left column, top to bottom*

Bright green thermophilic bacteria thrive in the mineral-rich hot water that flows in shallow channels along the terraces.

The microbial communities include *Chilodonella*, a microbial predator that eats diatoms. Diatoms have hard, glasslike shells. *Chilodonella* has tough rodlike "jaws" that pull entire diatoms into the cell (shown in the upper right of the photograph).

*Spirulina*, a spiral-shaped cyanobacterium, is one of the members of the brown mat communities found at Mammoth Hot Springs.

*Right column, top to bottom*

Photosynthetic bacteria have formed a green film on the surface of this small piece of travertine.

This piece of an orange mat was collected from the base of Orange Spring Mound. Many stringy bacteria grow together to form the mat.

Diatoms are present in this magnified picture of the orange mat.

Bacteria (some spiral, some not) and boat-shaped diatoms are intertwined in this magnified picture of an orange mat.

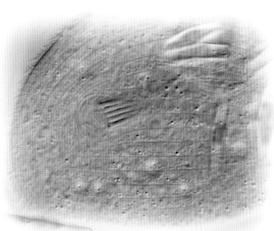
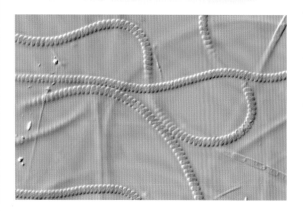
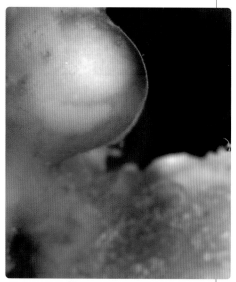

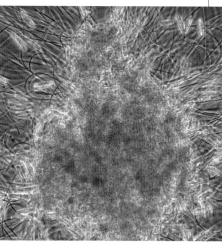
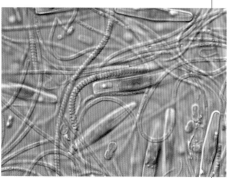

**Seen** *and* **Unseen**

## Angel Terrace

Angel Terrace is located on Upper Terrace Drive. Here, fast-flowing springs often rapidly create new formations. Bubbles of carbon dioxide gas are released. Blue-gray deposits of travertine form along the margins of the water flow. In some areas, the deposits inundate trees and foliage.

*Top* This view from the Upper Terrace Drive overlook at Mammoth Hot Springs is of a dramatic travertine terrace and a colorful blue spring. The brown and green colors in the runoff from the spring come from thermophilic microbes.

*Right* Brown microbial mats cover areas of Angel Terrace.

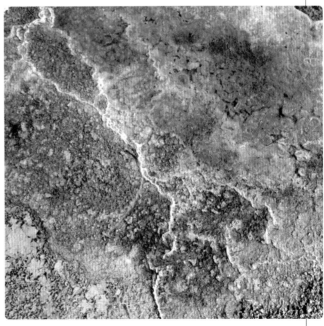

# Mammoth Hot Springs to Beaver Lake Area

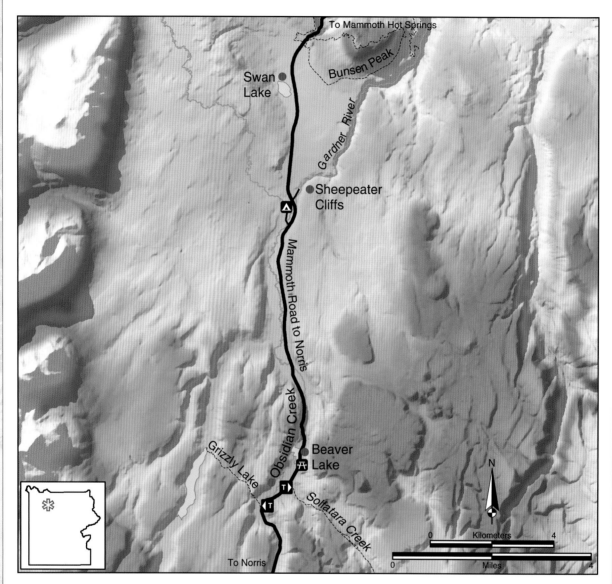

Map produced by the Spatial Analysis Center, Yellowstone National Park.

Picturesque Swan Lake is rich with microbes that provide nutrients, oxygen, and a healthy environment for the organisms that live in or visit its chilly water. Fine, muddy sediments along the shoreline collect debris from the nearby land, providing a rich habitat for microbes. Deep in the mud, oxygen is depleted, no sunlight is able to penetrate, and microbes adapted to low oxygen conditions thrive.

# Mammoth Hot Springs to Beaver Lake Area

## ● Swan Lake

Swan Lake is located in Swan Lake Flats, a high plateau southwest of Bunsen Peak. Temperatures are too cold for thermophiles, but the lake is rich with protozoa, algae, and bacterial communities adapted to a freshwater mountain lake. When the plants along the shore of the lake die, they provide organic matter that can be used as food by microbes. Protozoa are plentiful, feeding on **detritus** (dead and decaying debris), bacteria, algae, or each other. Algae produce oxygen that is utilized by other organisms in the lake. Microbes contribute to the food chain, recycling nutrients for other organisms, such as fish and birds. These microbial activities are crucial for maintaining a healthy ecosystem.

*Background* Plants along the lakeshore provide a sheltered habitat for microorganisms. Dead plant material, called detritus, that falls into the lake supplies nutrients.

*Inset top* Purple sulfur bacteria, found in the sediments of Swan Lake, capture sunlight to break apart hydrogen sulfide (the chemical that gives rotten eggs their distinctive smell) for energy. The sulfur deposits that give these cells a grainy appearance are by-products of this metabolism.

*Inset bottom* Pelomyxa is an amoeba that inhabits mud in freshwater sites where there is little or no oxygen. The bright bits inside the amoeba are pieces of sand.

*Right* Frontonia is common in freshwater lakes and streams. Just below the surface of the cell are hundreds of rod-shaped organelles called extrusomes. The extrusomes can be fired explosively from the cell for defense from other microscopic predators.

*Far right* Paramecium bursaria is green because hundreds of algae live inside it. The algae release oxygen as a by-product of photosynthesis. The oxygen is used by the Paramecia to survive in the low-oxygen mud on the lake bottom.

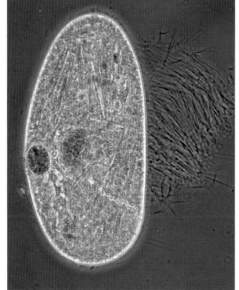
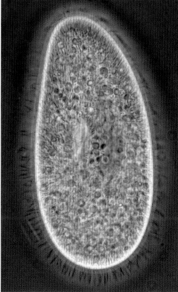

*Background* The Gardner River, near Sheepeater Cliffs, is swift and cold.

*Inset top Tabellaria* cells join end to end to form zigzag-shaped filaments. Millions of these common diatoms form orange or brown fluffy sheaths around submerged plants or rocks in cool-water habitats found worldwide.

*Inset middle Spirogyra*, a common, filamentous green alga in freshwater environments, has bright green chloroplasts that form a thin ribbon wrapped around the inside of the cell wall.

*Inset bottom Urostyla* is covered with specialized cilia, clustered together into little brushes called cirri. The cirri are strong enough to act as legs. The tips stick gently to surfaces allowing the *Urostyla* to "walk" over sand particles or pieces of detritus. Long, compound cilia, located at the front (left) of the cell, are used to sweep bacteria or algae into the mouth region.

## ● Gardner River at Sheepeater Cliffs

The Gardner River flows rapidly in a boulder-strewn streambed at the base of Sheepeater Cliffs. The water in this high-altitude mountain stream is not geothermally heated and is cold even in the summer months. Different kinds of microbes flourish at the water's edge in the nutrient-rich sediments trapped by rocks and debris. Other cold freshwater habitats throughout the world are home to similar types of microbes.

*The Habitats*

# Beaver Lake

Beaver Lake is near the Beaver Lake picnic area along the Mammoth to Norris Road just south of Obsidian Cliff. Water levels in this freshwater lake fluctuate seasonally depending upon rainfall and the spring snowmelt. In late summer, a scum layer forms in the shallow mudflats around the margins of the lake, providing excellent habitat for microbial communities. The top layer of the scum is home to many photosynthetic algae, especially diatoms. Diatoms are enclosed in a shell made from glassy siliceous material. The shells, known as **frustules,** do not decompose when diatoms die. Samples of mud from Beaver Lake contain many frustules from dead diatoms.

The mudflats also support many other kinds of algae, such as *Euglena*. Deeper in the mud layer, where sunlight does not penetrate and there is little oxygen, bacteria are more prevalent. Sometimes the bacteria produce by-products during their growth that can give the mud a sour smell.

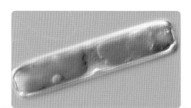

*Top to bottom*

This micrograph shows a sample of mud that is mostly composed of the shells of diatoms, one of the more common types of algae.

Diatoms have remarkably beautiful glassy shells called frustules. This empty frustule of a dead diatom has symmetrical ridges.

This picture of a live diatom shows the yellow-green chloroplasts and other parts of the cell enclosed within the frustule.

Beaver Lake, south of Obsidian Cliff, is surrounded by lodgepole pine and other conifers.

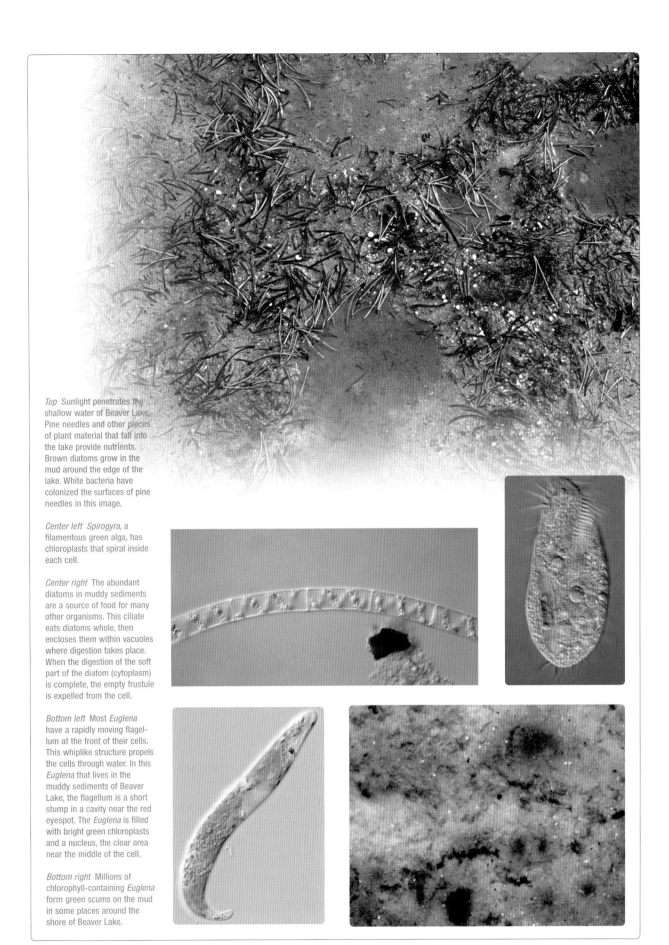

*Top* Sunlight penetrates the shallow water of Beaver Lake. Pine needles and other pieces of plant material that fall into the lake provide nutrients. Brown diatoms grow in the mud around the edge of the lake. White bacteria have colonized the surfaces of pine needles in this image.

*Center left* *Spirogyra*, a filamentous green alga, has chloroplasts that spiral inside each cell.

*Center right* The abundant diatoms in muddy sediments are a source of food for many other organisms. This ciliate eats diatoms whole, then encloses them within vacuoles where digestion takes place. When the digestion of the soft part of the diatom (cytoplasm) is complete, the empty frustule is expelled from the cell.

*Bottom left* Most *Euglena* have a rapidly moving flagellum at the front of their cells. This whiplike structure propels the cells through water. In this *Euglena* that lives in the muddy sediments of Beaver Lake, the flagellum is a short stump in a cavity near the red eyespot. The *Euglena* is filled with bright green chloroplasts and a nucleus, the clear area near the middle of the cell.

*Bottom right* Millions of chlorophyll-containing *Euglena* form green scums on the mud in some places around the shore of Beaver Lake.

# ● Obsidian Creek

Geothermally heated Obsidian Creek flows through a lush valley, 0.4 mile south of the Grizzly Lake trailhead. The water moves relatively slowly in the stream and traps dust and nutrient-rich debris along its surface. Communities of microbes form films or layers of mats in the debris. The water temperature and pH are ideal for the growth of cyanobacterial communities. *Oscillatoria*, a filamentous cyanobacterium, forms green regions in the mats. Iron bacteria grow in the red-colored areas. Other organisms such as **ciliates** (single-celled protozoa with hairlike projections called **cilia** that are used for movement or to capture food) attach to the mats and feed on the bacteria.

*Left* Photosynthetic bacteria live in the warm water of Obsidian Creek. Plants in the creek provide habitats that can be colonized by many microbes.

*Right, top to bottom*

Green microbial mats formed by communities of photosynthetic bacteria grow on the surface of the water. The thick mats collect dust and debris.

There are two types of filamentous cyanobacteria in this tiny piece of a mat collected from Obsidian Creek: a smaller and a larger species, one thin, the other much thicker.

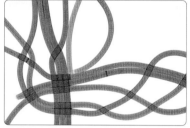

Bacteria containing red iron compounds give color to the mat at Obsidian Creek. This magnified piece of mat also contains some filamentous green cyanobacteria.

*Oscillatoria*, a cyanobacterium, has many cells joined end to end in long filaments that weave together to form large mats in Obsidian Creek.

This *Vorticella* is attached to the mat with a long stalk that moves by expanding and contracting. The *Vorticella* eats bacteria by sweeping them into its mouth.

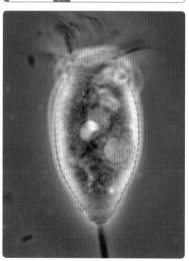

Lemonade Creek is covered in a bright green mat formed by *Cyanidium*, the most heat- and acid-tolerant alga known. *Cyanidium* can grow at pH as low as 1 and temperatures as high as 52°C (125.6°F). *Image provided by K. Sheehan.*

# Amphitheater Springs Area

The Amphitheater Springs Area is located along the Mammoth to Norris Road just south of the Beaver Lake picnic area near the Solfatara Creek Trail. In the Amphitheater Springs basin, steaming ground and hissing steam vents (fumaroles) create stark hillsides called solfataras. Large amounts of carbon dioxide, formed underground by volcanic activities, are released. Geological processes underground, as well as microbes that use sulfur compounds for metabolism, produce sulfuric acid and hydrogen sulfide gas (rotten eggs smell). The sulfuric acid creates features with very low pH values. Along these dry hillsides, what little water there is below ground eventually comes to the surface as steam, creating very hot soil temperatures. The soils appear barren and sterile, but some thermophilic microbes are able to survive even under these harsh conditions.

## Lemonade Creek

Lemonade Creek is a hot and acidic creek with temperatures as high as 60°C–70°C (140°F–158°F) and a pH of 2. Bright yellow sulfur deposits occur near the hot springs that feed the creek. As water flows away from the springs, it cools to about 50°C (122°F), allowing eukaryotic microbes such as acid-tolerant algae to grow. The most heat- and acid-tolerant alga known, *Cyanidium*, flourishes here in vivid green mats.

# Amphitheater Springs Area

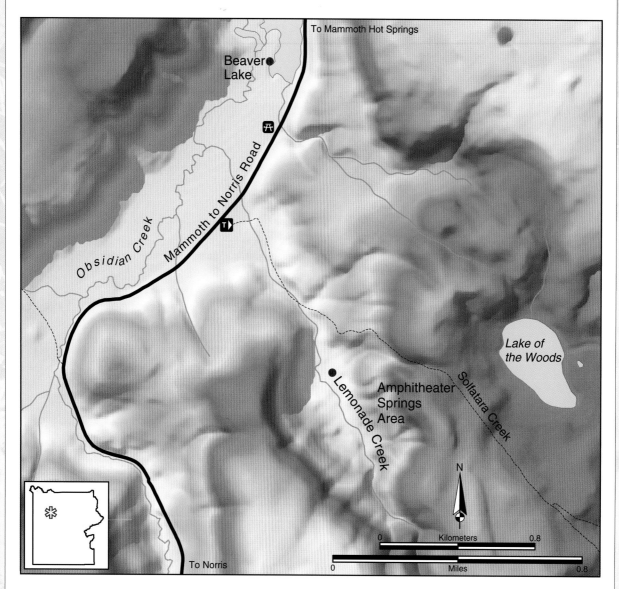

Map produced by the Spatial Analysis Center, Yellowstone National Park.

Springs like this support bacteria and archaea that utilize sulfur compounds for growth. Some bacteria contain deposits of pure sulfur inside. Other bacteria deposit sulfur outside their cells. Bright yellow sulfur deposits form in the hot spring or sometimes encrust microbes such as algae.

*Euglena*, adapted to very acid conditions, live in places in Lemonade Creek where temperatures are less than 45°C (113°F).

*Clockwise from top right* These highly magnified *Cyanidium* cells contain bright green photosynthetic pigments in their chloroplasts. Millions of these cells growing in mats in Lemonade Creek give it vivid color.

*Zygogonium* is a green alga whose cylinder-shaped cells join end to end in long strands.

*Zygogonium* grows at low pH in places in Lemonade Creek where the temperature is below 35°C (95°F). Sometimes the long filaments of *Zygogonium* are encrusted with sulfur.

This species of *Euglena* is adapted to the acidic conditions found in Lemonade Creek.

*Right inset* These two *Euglena* cells have bright red eyespots that sense light. The green regions contain the chloroplasts where photosynthesis takes place.

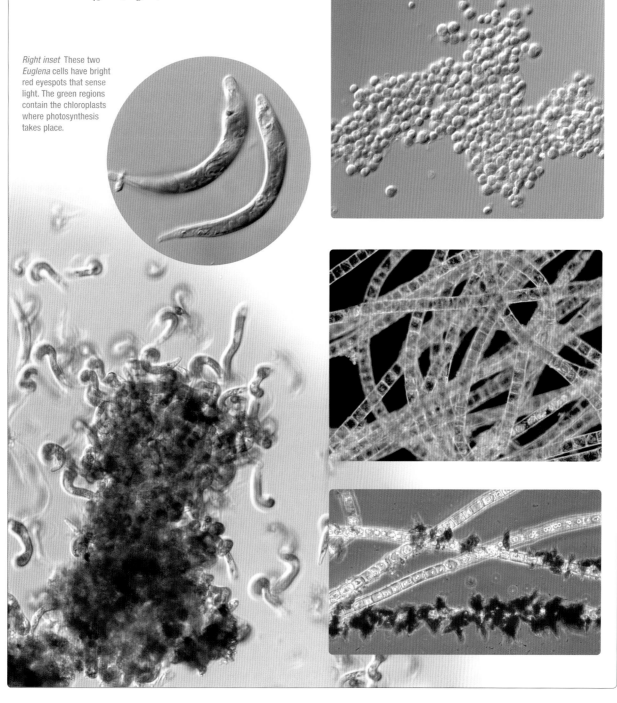

# Roaring Mountain to Norris Geyser Basin Area

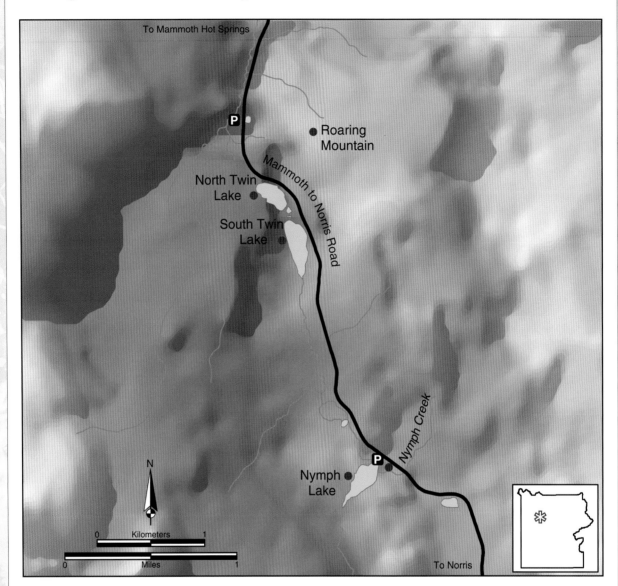

Map produced by the Spatial Analysis Center, Yellowstone National Park.

# Roaring Mountain to Norris Geyser Basin Area

## ● Roaring Mountain

Roaring Mountain is a barren hillside characterized by hot, steamy ground and fumaroles. The water and soil are highly acidic and rich in sulfur compounds, and the smell of hydrogen sulfide permeates the air. Vivid green *Cyanidium*, a thermophilic alga that grows at low pH, lives on rock surfaces and in moist soils in this area. Yellow sulfur deposits are visible around fumaroles. Throughout the area, sulfuric acid bleaches the color from the soil, creating stark features.

The steam rising from the fumaroles at Roaring Mountain creates a foreboding atmosphere. This area is characterized by very hot, acidic soils and a strong, rotten-egg smell.

# Twin Lakes

Twin Lakes (North and South) are located along the Mammoth to Norris Road south of Roaring Mountain. These cold-water lakes provide a tranquil contrast to Roaring Mountain's steaming and barren setting. Waterfowl are plentiful, and patches of water lilies grow along the shore of South Twin Lake. There are no thermophilic microbes here, but the lakes have an abundance of protozoa and algae typical of fresh-water lakes worldwide. Grasses growing along the shoreline provide a protected and nutrient-rich habitat for algae, protozoa, and bacteria.

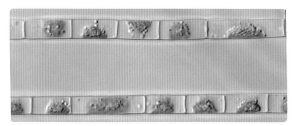

*Top left* The plants growing around the margins of the lake provide a rich habitat for diverse communities of microbes.

*Below* The verdant and tranquil shores of South Twin Lake provide a sharp contrast to the barren thermal areas at nearby Roaring Mountain.

*Above Klebsormidium* is a green alga that grows in South Twin Lake. It is composed of cells joined end to end. The chloroplasts are the bright green regions within the cell.

# Nymph Creek and Nymph Lake

Nymph Creek, a very hot and acidic creek, flows from springs near the Mammoth to Norris Road into the cooler water of Nymph Lake. Algae, adapted to high temperatures and low pH, carpet much of the streambed with a striking green mat. Other eukaryotic organisms, including *Euglena* and amoebae, live in the creek and lake. Photosynthetic bacteria like the cyanobacteria that live in other hot streams in Yellowstone cannot grow in the acidic water of Nymph Creek.

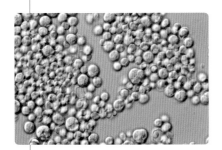

*Top left* Round *Cyanidium* cells with bright green chloroplasts.

*Center left* This is a small sample of the *Cyanidium* mat that covers much of the streambed at Nymph Creek.

*Bottom left* These are highly magnified *Hydrogenobaculum* filaments.

*Bottom center* These *Hydrogenobaculum* filaments are encrusted with sulfur.

*Bottom right* *Hydrogenobaculum* is a thermophilic bacterium with single cells joined end to end in long yellow streamers. It grows at 70°C (158°F) and pH 3 in Nymph Creek.

*Above* Nymph Creek is a hot, acidic creek located about a mile south of Roaring Mountain. Its most striking feature is a vivid green microbial mat dominated by algae.

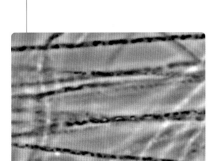
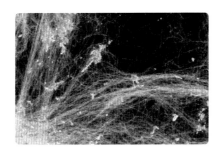

Layers of gold and brown diatoms color the mudflats along the shoreline of Nymph Lake. In areas where the water temperature is about 35°C (95°F), a filamentous alga, *Zygogonium,* grows in leathery, green or dark purple mats.

*Top* Nymph Lake has numerous thermal features along its shoreline.

*Left* In the cooler areas of Nymph Lake, *Zygogonium* forms purple leathery mats. Although *Zygogonium* tolerates very acid conditions, it cannot survive temperatures above about 35°C.

*Right top* The distinctive maroon color of *Zygogonium* comes from pigments present in the cell. The large green chloroplasts are visible in the center of each algal cell.

*Right center* *Eunotia* is a common diatom that often grows in large orange patches on the mud.

*Right bottom* This micrograph shows one type of the many species of diatoms found in the mud of Nymph Lake.

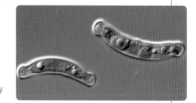

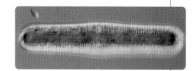

# Norris Geyser Basin

The Norris Geyser Basin, the hottest and most dynamic geothermal area in Yellowstone, contains many geysers, hot springs, and steam vents. Many features in the geyser basin are noted for their extremely acidic characteristics; among the most prominent of these is Echinus Geyser, the largest acidic geyser in the world. The basin includes alkaline springs and geysers as well, including Steamboat Geyser, the world's tallest geyser.

*Top* Porcelain Basin, located in the Norris Geyser Basin, has many striking geysers, hot springs, and fumaroles in one of the hottest and most active thermal areas in Yellowstone.

*Inset above* The colors in this hot-spring runoff channel are the result of iron-rich mineral deposits and microbes.

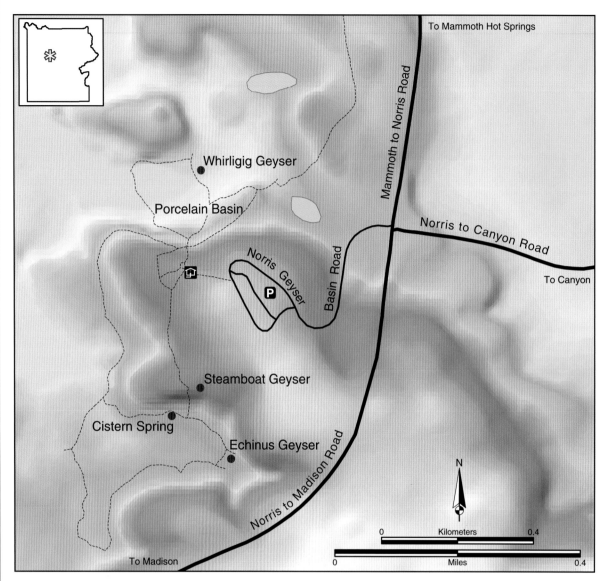

## Porcelain Basin

Porcelain Basin has hundreds of active thermal features that undergo dramatic changes. Geological processes such as earthquakes may cause some changes, and the underground water flow may influence activity. Many areas in the Porcelain Basin are hot and acidic and support the growth of microbes that thrive under these extreme conditions.

Different communities of microbes living along a striking temperature gradient create this dramatic pattern of colors in Porcelain Basin.

*The Habitats* 31

*Top* This mat in an acidic runoff channel in Porcelain Basin is colored bright green from chlorophyll found in the thermophilic alga *Cyanidium*.

*Inset* The colors around one of the hot, acidic outflows in Porcelain Basin are formed by microbial biofilms, reflected blue light from the mineral-laden water, and deposits of white sinter.

*Bottom* The terraces at Cistern Spring are formed as minerals from the constantly flowing water are rapidly deposited as sinter. The deposits surround and kill the lodgepole pines.

## ● Whirligig Geyser

Whirligig Geyser has an impressive, acidic runoff channel with bright colors created by mineral deposits and thermophiles. The bed of the channel combines yellow sulfur deposits with the bright green mat formed by the acidophilic alga *Cyanidium*.

## ● Cistern Spring

Cistern Spring, located along a boardwalk downslope from Steamboat Geyser, is connected to and affected by the activity of Steamboat Geyser. The spring, 300 feet away from Steamboat, sometimes drains completely following a major eruption of the geyser and takes three or four days to refill. It has a beautiful blue pool that spills over a series of sinter terraces. As the water flows over the terraces, it cools slightly and photosynthetic bacteria grow as colored **biofilms** (slimy substances that form on moist surfaces).

**Seen** *and* **Unseen**

# Lower Geyser Basin

## Fountain Paint Pot Area

The Fountain Paint Pot area is located in the Lower Geyser Basin. This area has numerous thermal features including mud pots, fumaroles, hot springs, and geysers. The mud pots are composed of mineral-rich clay and silica that are broken down into fine particles by the churning action of escaping steam and microbes that produce acid, which breaks down the underlying rock. The colors of the mud come from the composition of the underlying iron oxide–rich rocks.

*Above* Steam rises from the numerous thermal features found in the scenic Fountain Paint Pot area in the Lower Geyser Basin.

*Inset* Fountain Paint Pot is a mud pot with hot, acidic water that churns clay minerals and silica into fine, suspended particles.

*Bottom* Extensive microbial mats composed of heat-loving, photosynthetic bacteria colonize the outflow channels along the trails in the Fountain Paint Pot area. Cyanobacteria are able to live in alkaline water with temperatures as high as 74°C (165.2°F) and contain colored pigments that help protect them from damaging sunlight or aid in photosynthesis.

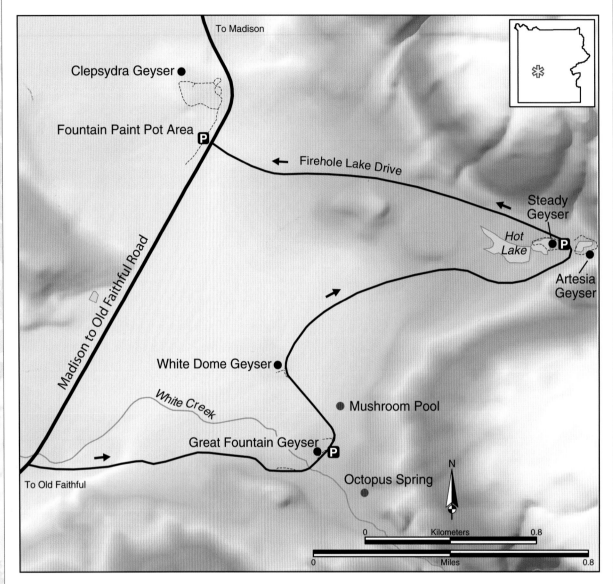

## ● Octopus Spring

Octopus Spring, located along White Creek near Firehole Lake Drive, is an alkaline spring that has a temperature of about 95°C (203°F). Microbiologists have intensely studied this spring for more than twenty years, identifying at least fifty-nine species of thermophilic microbes in the spring and its runoff channels. White sinter deposits form around the spring's edge. As the boiling water flows away from the spring, it cools and supports diverse communities of photosynthetic bacteria that form deep green and orange mats.

*Above* Octopus Spring is boiling hot at its source. As the water flows away from the main pool, it cools to temperatures that permit the growth of *Chloroflexus*, a photosynthetic cyanobacterium, and *Synechococcus*, a green nonsulfur bacterium, in the alkaline runoff channel.

*Inset* White crusty deposits form around the edges of Octopus Spring when mineral-rich steam evaporates.

*Bottom left* Thick microbial mats form in the outflow channels of the spring. These mats are composed primarily of diverse communities of cyanobacteria and green nonsulfur bacteria.

*Bottom right* A piece of the Octopus Spring mat was flattened and examined with a microscope. The curved rods with light-colored ends are *Synechococcus* cells. The filaments are the cyanobacterium *Chloroflexus*. Although these species are dominant in the mat, other bacteria also are present in this image.

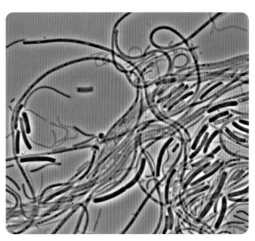

# Mushroom Pool

Mushroom Pool is located along Firehole Lake Drive. The spring has a significant place in the history of microbiological research in Yellowstone. Thomas Brock discovered *Thermus aquaticus,* the bacterium that was the first source of *Taq* polymerase, in samples collected from Mushroom Pool. At that time, microbiologists had no idea that life could survive at temperatures as high as 80°C (176°F). Brock's discovery led to the search for other species that grow at high temperatures.

The thermophilic bacterium *Thermus aquaticus* was first isolated in Mushroom Pool.

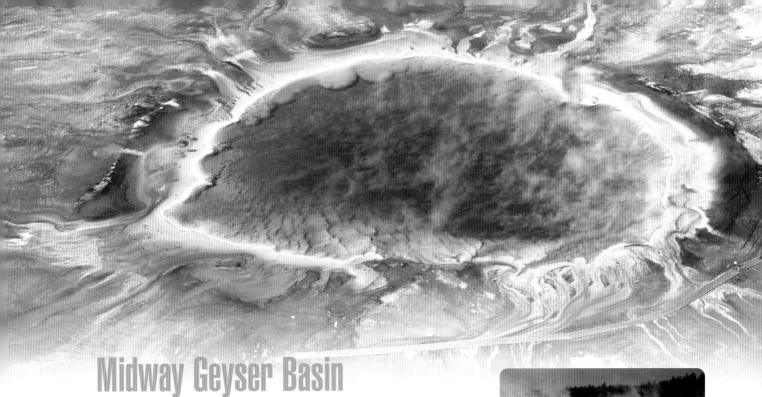

# Midway Geyser Basin

## ● Grand Prismatic Spring

Grand Prismatic Spring is in the Midway Geyser Basin near the Firehole River. This brilliantly colored hot spring is 370 feet in diameter, making it the largest hot spring in Yellowstone and the third largest in the world. Its temperature ranges from 63°C to 87°C (145.4°F to 188.6°F). Where temperatures are cool enough to permit (about 70°C or 158°F), photosynthetic bacteria form extensive, spectacularly colored mats ranging in color from pale green and yellow to dark red and orange.

Massive Excelsior Geyser releases approximately four thousand gallons of water each minute. As the hot runoff flows away from the geyser into the adjacent Firehole River, it cools and forms thermal **gradients** (distinct zones containing communities of microbes adapted to specific temperatures) rich with photosynthetic bacteria. As the hot runoff flows away from the geyser into the adjacent Firehole River, it cools. Communities of microbes, adapted to specific temperatures, grow in these runoff channels and their pigments create characteristic bands of color. The colors can vary during

*Above* This aerial view of spectacular Grand Prismatic Spring, located in the Midway Geyser Basin, shows extensive orange and brown microbial mats radiating from the pool. Rainbow-colored steam often rises from the surface of the nearly boiling water in the middle of the hot spring. Only hyperthermophiles are able to survive in water this hot. *Image courtesy of the National Park Service.*

*Right* The walkways at Grand Prismatic Spring allow visitors to view rare and extraordinary microbial mats.

different seasons of the year depending upon the amount of sunlight falling on the mats. This creates a light gradient. The photosynthetic microbes at the top of a mat produce deep red and orange pigments for protection from intense sunlight during the summer. During the winter months when less light penetrates the mat, the green photosynthetic pigments predominate. Some cyanobacteria can move to places in the mat with light and temperature conditions favorable for their growth requirements.

*The Habitats*

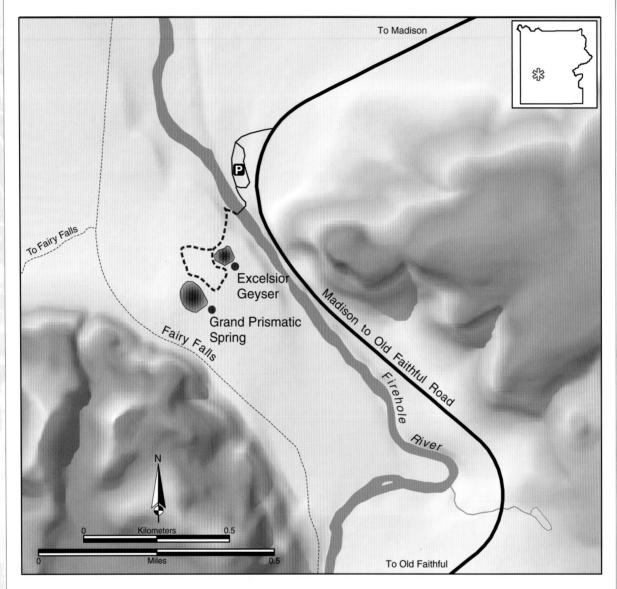

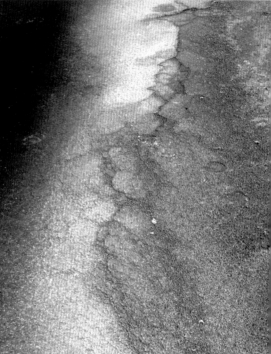

*Above* Layers of silica-rich minerals are deposited in the outflows from the springs and are colonized by microbial communities that form beautiful patterns.

*Inset above* In winter, hot steam from Excelsior Geyser condenses in the cold air, creating an eerie landscape.

*Top right* Microbes containing brown pigments live in the outer edges of the microbial mats. When boiling water flows away from the hot spring, it cools enough to allow photosynthetic bacteria, such as *Chloroflexus* and *Synechococcus*, to colonize the channels.

*Center right* Mineral deposits and microbes are intimately linked to create these colorful formations.

*Bottom right* Millions of gallons of hot water pour from Excelsior Geyser into the Firehole River each day. Thermophilic microbes thrive in the outflow.

# Upper Geyser Basin

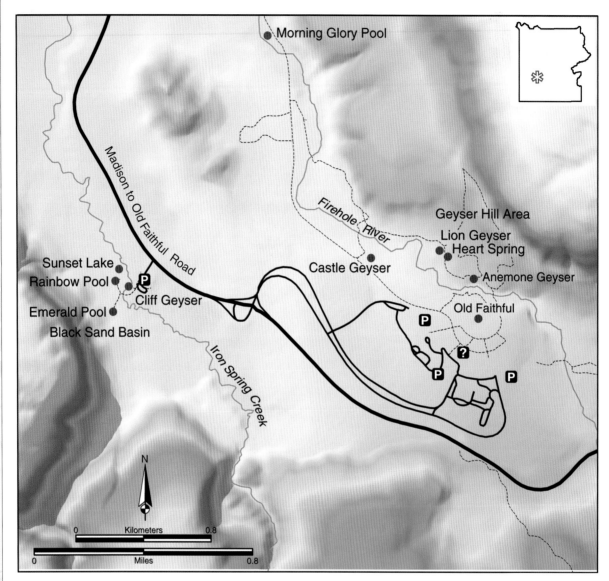

# Upper Geyser Basin

## ● Black Sand Basin

Black Sand Basin, an intensely beautiful area of hot springs in the Upper Geyser Basin, contains several brilliantly colored features, including Emerald Pool, Rainbow Pool, and Sunset Lake. Photosynthetic bacteria grow in the outflows into Iron Spring Creek, especially in the vicinity of Cliff Geyser, creating strikingly beautiful colors and extensive microbial mats.

Emerald Pool has a deep green color, caused by the reflection of light from a combination of the blue color of the silica-rich water and yellow-colored sulfur deposits and communities of photosynthetic bacteria.

Sunset Lake is one of the largest active thermal features in Yellowstone. Small eruptions occur frequently. The resulting runoff channels host beautifully colored microbial communities.

*Above* Emerald Pool is one of the striking hot springs located in Black Sand Basin.

*Below* The bands of color in this outflow channel are created by different bacterial communities living at different water temperatures. Cyanobacteria containing orange and brown pigments live in a narrow temperature range. Other green photosynthetic bacteria are present at slightly cooler temperatures.

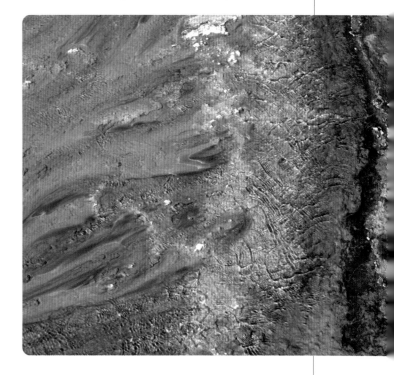

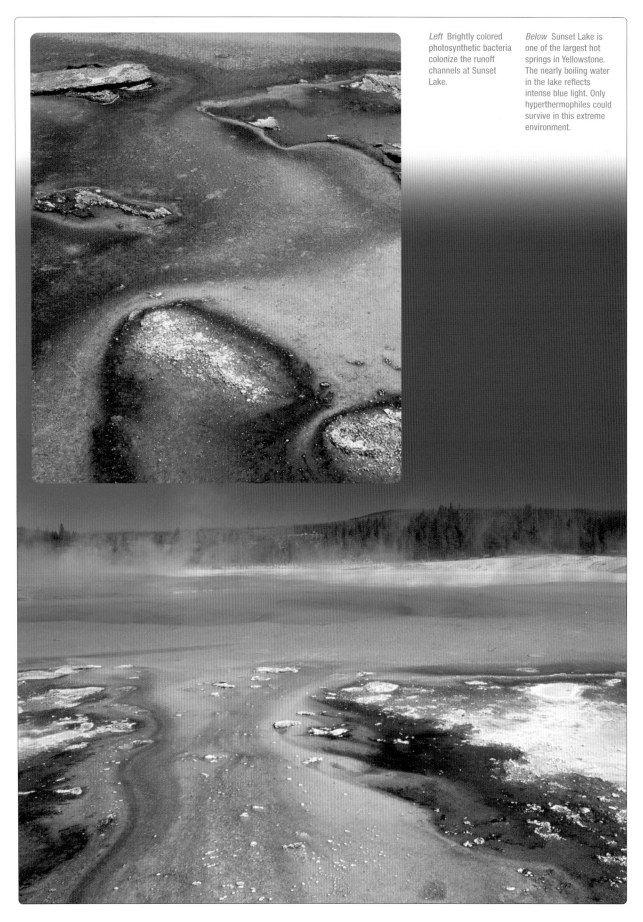

*Left* Brightly colored photosynthetic bacteria colonize the runoff channels at Sunset Lake.

*Below* Sunset Lake is one of the largest hot springs in Yellowstone. The nearly boiling water in the lake reflects intense blue light. Only hyperthermophiles could survive in this extreme environment.

*Background* Old Faithful, one of the most famous and photographed geysers in the world, is an icon of Yellowstone National Park.

*Inset above* Geyser Hill is located across the Firehole River from Old Faithful Geyser. More than thirty-five geysers are located in this area.

*Below* Along the Firehole River, extensive sinter deposits and mineral crusts form steep banks that are colonized by communities of thermophilic bacteria.

## Old Faithful Area

The Old Faithful area is home to the Upper Geyser Basin and contains the world's largest concentration of geysers. Geyser Hill, located across the Firehole River from Old Faithful Geyser, is the site of some of the largest and most active geysers found anywhere in the world.

When superheated water from underground that is under pressure and exceeds the boiling point for water (92.7°C–93°C or 199°F–199.4°F in Yellowstone) is expelled at the surface, minerals precipitate and form gray, silica-rich deposits called sinter. Large expanses of sinter deposits exist on Geyser Hill, especially near Heart Spring. Thermal runoff flows from the features on Geyser Hill into the Firehole River. These outflows are relatively cool enough (74°C or 165.2°F), compared with boiling water expelled from the geysers, to permit the growth of brightly colored photosynthetic bacteria. These communities of bacteria are present in some places along the riverbank, around the perimeters of hot springs, and in the runoff from the thermal features.

Castle Geyser erupts from an impressive sinter cone. The extremely hot water emitted from the geyser flows away from the cone and cools in the runoff channels. Photosynthetic microbes colonize these areas and form colored biofilms.

Morning Glory Pool is located almost midway between the Geyser Hill area and Biscuit Basin to the north. Money, clothing, rocks, and other debris thrown into it by vandals clogged the pool and decreased its water temperature and flow. This permitted the growth of cyanobacteria and changed the color of the pool from intense, radiant blue to the present-day orange and yellow.

*Above* Castle Geyser is one of the largest sinter formations in the world. The geyser erupts about every ten to twelve hours, and eruptions often reach 90 feet tall. Colored microbial mat communities thrive in the outflow from the geyser.

*Top right* Heart Spring, located near the Lion Geyser group on Geyser Hill, contains silica-rich water that reflects brilliant blue light. Sinter deposits form along its edges.

*Right* Anemone Geyser, located on Geyser Hill, erupts every seven to fifteen minutes. Communities of photosynthetic bacteria live around the edges of the sinter-encrusted pool.

*Below* Remarkable Morning Glory Pool is encrusted with colorful mineral deposits and bacteria.

# West Thumb Geyser Basin

The West Thumb Geyser Basin lies along the western shores of Yellowstone Lake. A volcanic eruption about 160,000 years ago created a crater that is the West Thumb of Yellowstone Lake. The basin contains many active hot springs, pouring thousands of gallons of hot water into the lake each day. Where temperatures allow the growth of communities of thermophilic bacteria, colorful mats form in the outflows or along the margins of the pools and mud pots. Some of the thermal features extend into Yellowstone Lake.

## Thumb Paint Pots

The Thumb Paint Pots contain hot, acidic water that dissolves the underlying rocks into fine particles of mud. Thermophiles live along the edges of the pools. Small areas around the margins of the thermal features, where the pH is neutral to alkaline, are populated with dark green communities of cyanobacteria.

These hot pools located along the boardwalk near the Thumb Paint Pots contain pink- and orange-pigmented microbes around their edges.

*The Habitats*

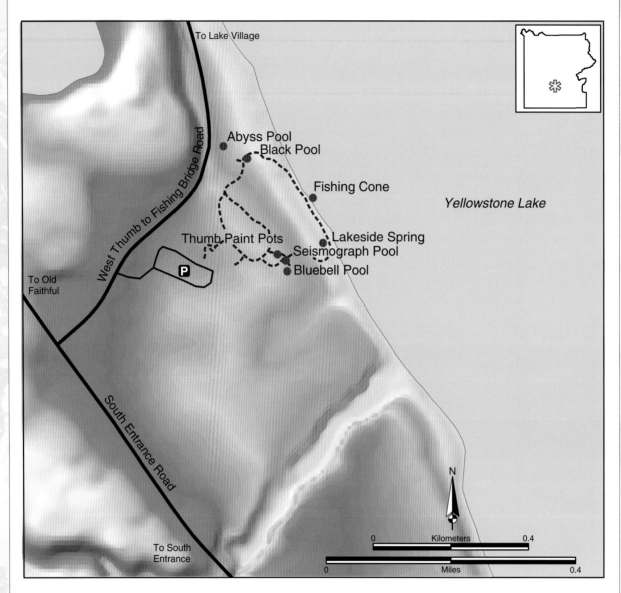

# Seismograph and Bluebell Pools

These two pools have temperatures close to boiling, preventing the growth of photosynthetic bacteria except along the margins of the pools where the water temperature is about 70°C (158°F). The water in the pools absorbs all wavelengths of sunlight except blue, which is refracted and made visible, giving the features the appearance of blue.

*Below* Near Seismograph and Bluebell Pools, a tiny hole along the boardwalk teems with microbes. The dark olive color is one of the characteristic colors of cyanobacteria.

*Right* This jellylike clump of a bacterial mat is composed of long filaments formed by photosynthetic cyanobacteria.

*The Habitats*

# Lakeside Spring

Lakeside Spring flows downhill into Yellowstone Lake. Extensive orange, red, and green microbial mats form along the hillside and are easily viewed from the boardwalks.

*Below* Extensive orange and green mats are found in the runoff channels from Lakeside Spring on the shore of Yellowstone Lake.

*Inset* A piece of the green-pigmented bacterial mat has a loose, fluffy composition.

*Above* The mat contains *Synechococcus,* a curved, sausage-shaped bacterium.

48    Seen *and* Unseen

## Fishing Cone

Fishing Cone, located offshore in Yellowstone Lake, is one of the most distinctive thermal features in the park. The large, steaming sinter cone is sometimes covered by water when the lake's water level is high. Photosynthetic bacteria grow along one side where heat and moisture conditions are favorable for growth. Early visitors to Yellowstone claimed they could catch fish and then cook them in the hot water in Fishing Cone.

## Black Pool

Black Pool at West Thumb Geyser Basin was named when the pool was cooler than it is today and supported large numbers of dark green cyanobacteria. In the early 1990s, the pool's temperature increased dramatically and the cyanobacteria were killed. Today the pool is an intense blue color. Photosynthetic bacteria are able to grow only along the relatively cooler edges of the pool. Now microbial mats form only along its margins. In places, tiny vertical **stromatolites** are present. Stromatolites are layered structures formed when cyanobacteria and photosynthetic bacteria grow together in mats that trap sand and organic material. In some kinds of stromatolites, the layers can harden into fossils over a long period of time. Fossil stromatolites that are about 3.5 billion years old provide evidence of the earliest life on Earth.

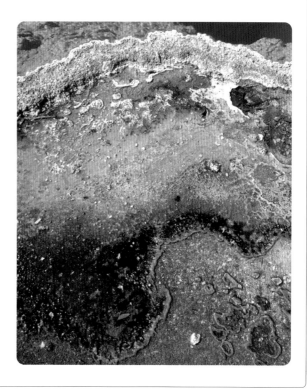

*Top* Even the famous Fishing Cone on the shore of Yellowstone Lake has communities of bacteria colonizing its rim.

*Right* The edge of Black Pool has crusty white mineral deposits and zones of orange, green, and brown formed by the pigments in photosynthetic bacteria.

# Abyss Pool

Abyss Pool once had periods of violent eruptions. Now, dark brown mats, typical of mats found in the West Thumb Geyser Basin, surround the intense turquoise-blue pool. The pool is 53 feet deep and is one of the deepest hot springs in the park.

*Background* The upper temperature limit for photosynthesis is about 74°C (165.2°F). Photosynthetic cyanobacteria are able to survive in Abyss Pool when the runoff cools along the edges.

*Inset top left* Long stringy filaments of cyanobacteria are intertwined to form the brown mat.

*Inset left* A closer microscopic view of the brown mat shows a tangle of filamentous bacteria.

*Below right* The orange and brown region around the edge of Abyss Pool at West Thumb Geyser Basin was created by microbes.

The Mud Volcano region contains bizarre formations. The air is filled with the strong smell of hydrogen sulfide gas. Heat-loving microbes, called *Sulfolobus*, that resemble some of the earliest organisms on Earth (Archaea) live in the sulfur-rich pools. Many microbes use sulfur for growth and produce sulfuric acid as a waste product. The acid dissolves the rocks into fine mud particles.

# Mud Volcano Area

The Mud Volcano area is located along the Yellowstone River between Fishing Bridge and Canyon. Hissing and violently churning mud pots provide evidence of the intense geothermal activity below the surface. Minor earthquakes are common, and a strong odor of hydrogen sulfide gas pervades.

The Mud Volcano area is very hot and acidic, and extensive bacterial mats do not form here. Some hyperthermophiles use the abundant sulfur compounds as energy sources, producing sulfuric acid as a by-product. This acid and the churning action of the hot water breaks down the underlying rock into fine, suspended clay particles, making the water muddy.

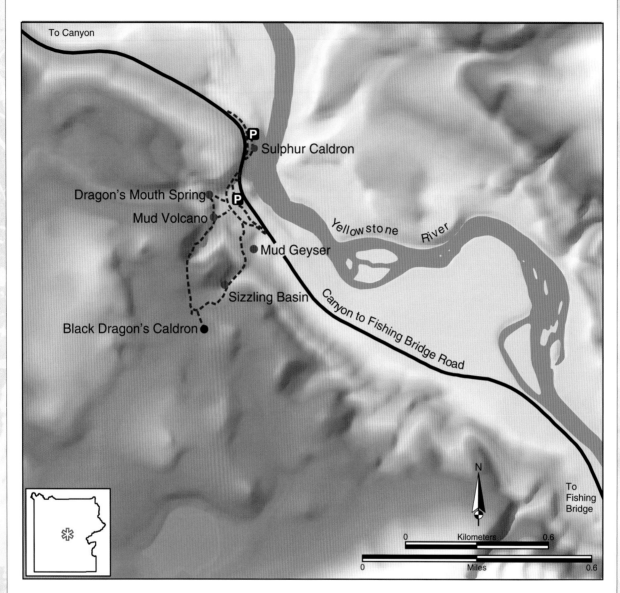

# Mud Geyser and Sizzling Basin

Frequent earthquake activity influences the soil and water temperatures of Mud Geyser and Sizzling Basin. The steaming pools contain fine rock particles in suspension. Although the whole area seems inhospitable, thermophilic microbes live in the mineral-rich water and along the shallow edges of the pools.

*Background* Green microbial mats, composed primarily of acid-tolerant algae, grow in the relatively cooler regions along the edges of the boiling pools in the Sizzling Basin area.

*Inset* The hot, acidic water in Mud Geyser dissolves rocks into fine particles that are suspended in the water to give the pool an opaque appearance.

# Dragon's Mouth Spring and Sulphur Caldron

These belching and hissing springs of churning chalk-colored mud may look deadly to the human eye, but nevertheless they contain hyperthermophilic archaea that thrive in this hot, sulfur-rich environment. Other thermophilic bacteria and acid-loving algae form colored bands along the edges of Dragon's Mouth Spring. Sulphur Caldron, located across the road from the Mud Volcano area, is one of the most acidic thermal features in Yellowstone, with a pH similar to stomach acid.

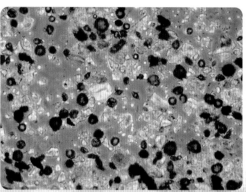

*Top right* Dragon's Mouth belches steam and gases from a hillside cavern. Despite the extremely hot (80°C or 176°F) and acidic (pH 1.2) water, hyperthermophiles, specifically *Sulfolobus*, thrive in this harsh environment.

*Right* The water in Dragon's Mouth contains fine, suspended soil particles.

*Below* Only hyperthermophiles like *Sulfolobus* can grow in the constantly churning Sulphur Caldron. *Image provided by the National Park Service.*

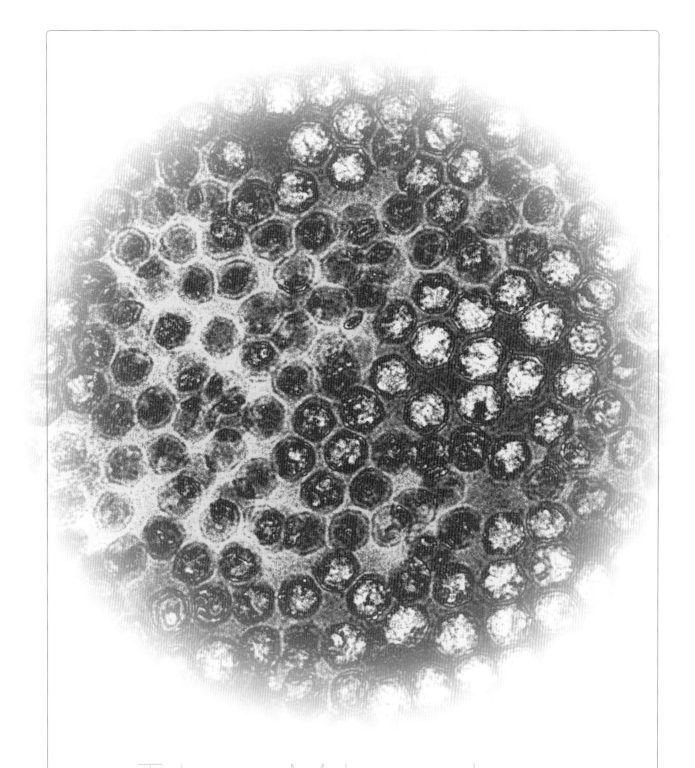

The Microbes

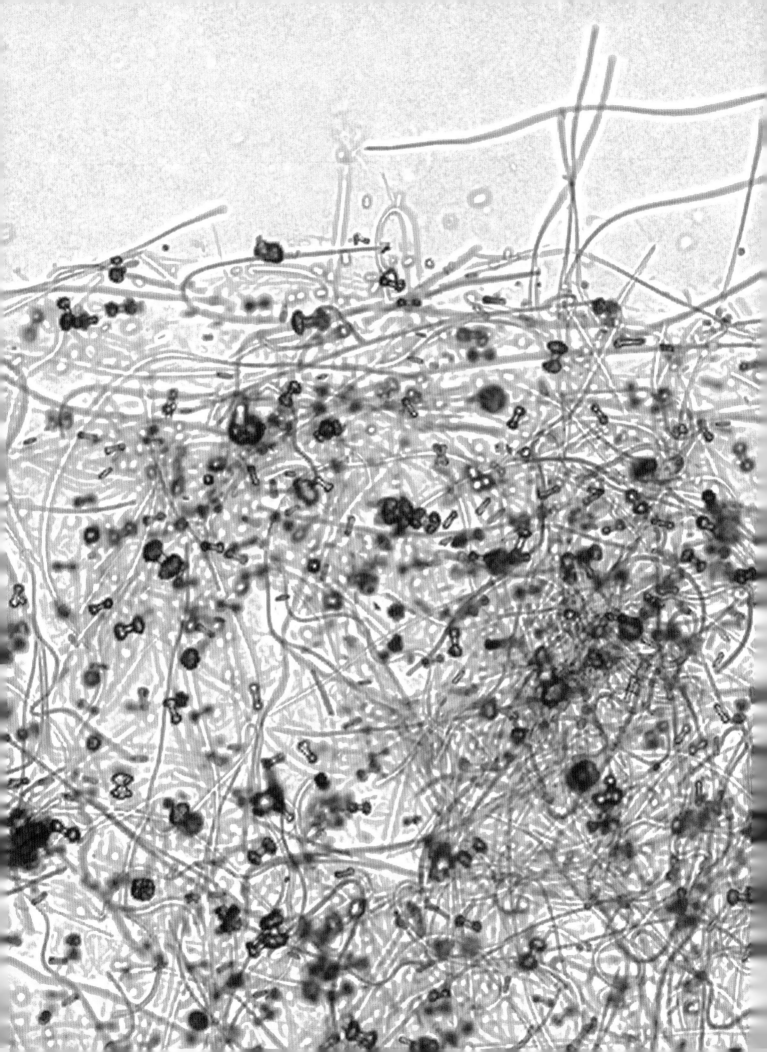

Biologists classify all organisms on Earth into three broad categories called **domains:** Archaea (sometimes called Archaebacteria), Bacteria (sometimes called Eubacteria), and Eukarya. The relationships of the organisms within the domains and their evolutionary history (ancestry) are displayed in diagrams often called "trees of life." The root of a tree represents the common ancestor of all life. The trunks correspond to the three domains, and the branches signify the major groups of organisms. Relationships are determined by studying fossil materials and by comparing living organisms for similarities and differences between the molecules found in them.

Studies that compare the makeup of genetic material have played an important role in developing the current understanding of the evolution of life. These genes are present in all living organisms, from microbes to higher plants and animals. As the genes are passed from one generation to the next, changes occur in the genetic code of the genes, and the accumulation of such changes provides evidence about the evolutionary history of organisms. Simply stated, the more similar the makeup of the genes, the more related the organisms are, and the closer they are grouped on a tree of life. Modern techniques in biology—including the development of polymerase chain reaction or PCR (a method of copying DNA), the ability to determine the genetic code, and sophisticated computerized databases containing millions of gene sequences—have made comparisons routine for biologists.

One revolutionary finding in the late 1970s changed how scientists classified organisms. Up to that time, biologists divided organisms into two groups, the Prokaryotes (or Bacteria) and the Eukaryotes. Analyses of the genetic material, called **ribosomal RNA,** revealed that a major new group of microbes, the Archaea,

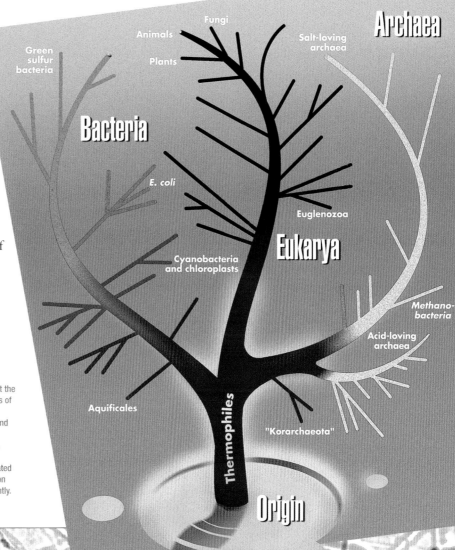

This diagram shows the relationships between all major forms of life based upon comparisons of one particular molecule, the gene for the small subunit of ribosomal RNA. The tree provides insights into the evolutionary history of life on Earth and shows that the three major branches of life are interrelated. Microbes (Archaea and Bacteria) form two branches of the tree. Animals, plants, and fungi are closely related and have appeared on Earth relatively recently.

previously classified within the Bacteria, were fundamentally different from the other prokaryotes. Although archaeal microbes are similar in size and appearance to bacteria, comparisons of DNA led to the conclusion that archaea represent a previously unrecognized category of organisms as distinct from bacteria as humans are. They now form a major trunk on the tree of life.

The lengths of the lines of evolutionary trees are used to depict the number of random changes (**mutations**) that have occurred during an organism's history. Archaea have short lines that often branch near the bottom of the evolutionary tree. This suggests that the archaea have not evolved, or mutated, as rapidly as other organisms on Earth. Scientists speculate that the harsh environments where archaea live limit the amount of genetic change that can occur in these organisms and still permit their survival. If this is true, the hyperthermophilic archaea, like ones found in Yellowstone, could be very similar to the earliest microbes on Earth.

Microbes, like other organisms, evolve as a result of mutations in their genetic makeup. The processes of natural selection eliminate mutations that make organisms less viable or fit. Microbes reproduce quickly and thus evolutionary changes can occur rapidly. Many bacteria also have the ability to acquire genes from other organisms. Some of the genes impart beneficial properties to the bacteria. This process allows bacteria to adapt quickly to physical, chemical, and biological changes in their environments. The ability of bacteria to evolve drug-resistant strains is an example of this process.

Temperature may be one of the most important variables in the growth of organisms, because most organisms are able to survive only within a narrow temperature range. Humans feel pain when immersing their hands in water at 50°C (122°F). Some eukaryotic microbes can grow at temperatures as high as 60°C–62°C (140°F–143.6°F). The photosynthetic cyanobacteria can live at 74°C (165.2°F). At higher temperatures, up to about 90°C (194°F), nonphotosynthetic bacteria are able to grow. In Yellowstone, the remarkable hyperthermophilic archaea thrive at temperatures above the boiling point for water (99°C or 210.2°F).

The presence or absence of oxygen in an environment also greatly influences microbial growth. Many microbes live where there is little or no oxygen. Even a brief exposure to air will kill some, whereas others can tolerate short exposures to air. For most of its history, Earth was devoid of oxygen, and early microbes derived energy from other abundant gases found in the atmosphere, such as carbon dioxide. Many of the hyperthermophilic archaea present on Earth today use gases such as hydrogen for energy. These microbes provide clues about what life might have been like several billion years ago.

## Upper Temperature LIMITS FOR LIFE

**Hyperthermophiles**

| | |
|---|---|
| Archaea | 113°C (235.4°F) |

**Thermophiles**

| | |
|---|---|
| Bacteria | 90°C (194°F) |
| Cyanobacteria | 74°C (165.2°F) |
| Fungi | 62°C (143.6°F) |
| Algae | 60°C (140°F) |
| Protozoa | 56°C (132.8°F) |
| Mosses, Insects, Crustaceans | 50°C (122°F) |
| Higher Plants | 45°C (113°F) |
| Fish | 38°C (100.4°F) |

All microbes require water, but many organisms are adapted to growth and survival with very little water. Some bacteria and fungi produce spores with cell walls that are resistant to desiccation (drying). The spores can exist in a dormant state until water becomes available for growth. Other microbes can dry out, then rejuvenate when water is available. Organisms such as lichens grow slowly and many are uniquely adapted to extremely dry environments.

Living organisms display a kaleidoscope of colors, from deep blue columbine flowers to vivid pink flamingoes. The colors come from natural pigments (colored chemical compounds) found in cells. The pigments carry out many functions. In humans, the pigment melanin can act as a sunscreen to protect skin from damaging light. Green chlorophyll and **carotenoids** important for photosynthesis are common in plants, algae, and photosynthetic bacteria. Carotenoids are pigments that give plant leaves, fruits, flowers, and vegetables their orange, red, and yellow hues. Carrots, for example, are orange because of carotenoids.

Carotenoids capture the wavelengths of sunlight that are not absorbed by the chlorophylls during photosynthesis. Usually the plentiful green chlorophylls hide the colors of the carotenoids, but sometimes their orange, yellow, red, and brown colors are visible in places like the cyanobacterial mats in Yellowstone or in autumn leaves.

Other bacteria, such as the purple bacteria and green bacteria, have different types of chlorophylls that absorb light at other wavelengths. These microbes often live together, exploiting the wavelengths of light that are unused by the other organisms in the habitat.

Colors in microbial mats can change depending upon the amount of sunlight. In summer, when there is lots of sunlight, carotenoids are more prominent and mask the green chlorophylls found in the bacteria. When sunlight is low, such as in the winter or when several cloudy days occur, the green chlorophylls predominate, giving the otherwise brilliant colors of the mats a more muted appearance.

This diatom contains brown photosynthetic pigments within organelles called chloroplasts.

## Viruses

Viruses are not like living cells. A virus "particle" requires a living host cell in order to reproduce. The virus inserts its DNA or RNA into the host cell. Sometimes this gives the host cell beneficial properties that can be inherited in the next generation of cells. But, more often, a virus kills the host cell and is released back into the environment. A protein coat or shell protects the genetic material in the virus until a new host cell is encountered. Sometimes viruses remain in a dormant state inside a host cell, reappearing later and beginning a new cycle.

Although people usually think of viruses as the cause of diseases, viruses also are an integral part of natural ecosystems. They are predators of other microbes and keep populations in balance. Viruses that infect bacteria are called **bacteriophages.**

Viruses that multiply inside thermophilic bacteria are especially interesting to scientists because they provide clues about high-temperature biochemistry. Recently, viruses have been isolated from *Sulfolobus*, a hyperthermophilic archaeal organism found in hot springs in the Norris Geyser Basin and Mud Volcano area. Some of the Yellowstone viruses are similar to ones isolated from *Sulfolobus* found

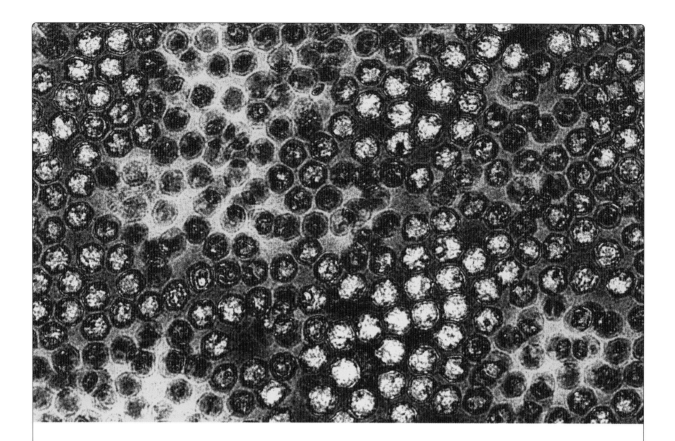

This electron microscope image shows many virus particles isolated from *Sulfolobus*, a hyperthermophile that lives in very hot and acidic springs. *Image provided by M. Young and G. Rice and color enhanced by D. Patterson.*

in different thermal areas of the world. Others are completely new.

# Archaea

The archaea include the most heat-tolerant microbes, and they grow at the most extreme temperatures of any known organisms on Earth. Archaea are diverse and are found in various habitats from hot springs to cold oceans to salty inland seas.

The Korarchaeota are a large group of hyperthermophiles that were discovered in Obsidian Pool in Yellowstone. These microbes may be most similar to the earliest living organisms on Earth.

## Sulfolobus

*Sulfolobus* thrives in a witch's brew of boiling acid. It is an extreme thermophile, growing without oxygen at temperatures as high as 90°C (194°F) in sulfur-rich hot springs. It also is an extreme acidophile, preferring growth at pH 2 or 3. *Sulfolobus* derives energy by metabolizing pure sulfur or sulfur compounds, such as hydrogen sulfide, into sulfuric acid.

# Bacteria

There are enormous numbers and types of bacteria due in part to their long evolutionary history on

These *Sulfolobus* cells were photographed using a transmission electron microscope. The grainy region inside the cells is the DNA. *Image provided by M. Young and S. Brumfield and color enhanced by D. Patterson.*

Earth, their ability to exploit many chemical compounds or sunlight as energy sources, and their tiny size.

## ● Hydrogenobaculum

*Hydrogenobaculum* belongs to the Aquificales, a group of bacteria that emerged early during the evolution of life. *Hydrogenobaculum* likes it hot. In Yellowstone, large numbers of *Hydrogenobaculum* cells often intertwine to form yellow and white bacterial streamers, or long hairlike filaments. Sometimes the filaments contain crystals of sulfur, which give them a creamy color. *Hydrogenobaculum* also has a sausage-shaped swimming form. *Hydrogenobaculum* derives energy from hydrogen and grows at temperatures above 65°C (149°F). Earth's early atmosphere had abundant hydrogen, and organisms like *Hydrogenobaculum* may resemble some ancestral microbes.

*Background* *Hydrogenobaculum* and other thermophiles grow in thin white films on the terraces at Mammoth Hot Springs. The delicate and beautiful films are called streamers.

*Top right* Under the microscope, mineral deposits are often observed on the streamers.

*Center right* Long, thin filaments formed by bacteria such as *Hydrogenobaculum* compose the wispy streamers.

*Bottom right* When microbiologists grow the bacteria found in the streamer, the rod-shaped cells predominate, illustrating that cells grown in a laboratory don't always resemble the streamers found on the terraces. *Image provided by J. Donohoe-Christiansen.*

## Thermus aquaticus

*Thermus aquaticus* is a thermophilic, rod-shaped bacterium that sometimes forms streamers. It grows ideally at neutral pH in habitats with temperatures between 40°C and 79°C (104°F and 174.2°F). It was first isolated from Yellowstone, but since then it has been found in many thermal environments, including hot-water heaters. The discovery of this bacterium growing in a hot spring was important: It led to the exploration of life at high temperatures and the discovery that there are many other thermophilic bacteria.

*Thermus aquaticus* was the original source of *Taq* polymerase. Since the initial isolation of *Taq* polymerase, researchers have discovered other heat-stable polymerases. For example, vent polymerase was isolated from an archaeal species of *Thermococcus* found in deep-sea hydrothermal vents. Often, *Thermus* forms bright red- or orange-pigmented filaments or yellow-pigmented colonies. The carotenoid pigments protect the bacteria from high levels of sunlight.

## Iron-Oxidizing Bacteria

Many thermal areas and hot springs in Yellowstone have high concentrations of iron. Iron-oxidizing bacteria use iron compounds to metabolize energy for growth. During this growth, iron compounds, which are insoluble in water, form and precipitate as bright orange rustlike deposits.

Scientists have shown that the orange-brown mat at Dragon Springs, a research study site in Norris Geyser Basin, not only contains iron, but also contains arsenic concentrations that are toxic for most organisms. Bacteria adapted to high levels of arsenic in these orange mats may be utilizing arsenic to provide energy, much as the iron-oxidizing bacteria use iron. There are many different kinds of iron bacteria. Some are filaments and some are small spherical cells or short rods. Most produce mucus material containing iron that surrounds the cells. Millions of iron bacterial cells growing together produce red, rustlike deposits that contain enough iron to make the cells brittle.

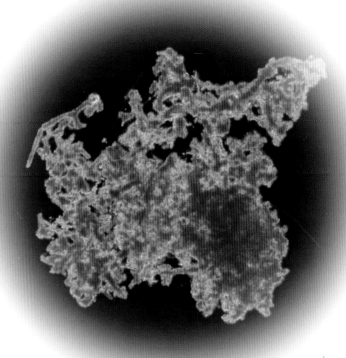

*Above* The bacterium *Gallionella*, found in La Duke Spring, secretes an iron-rich mucus that forms underwater films. This film contains enough iron to make it brittle.

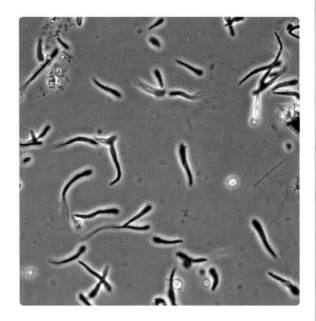

*Below* An iron-rich mucus coating surrounds the individual bacterial cells.

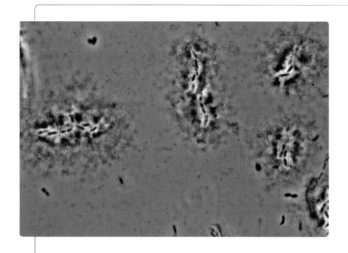

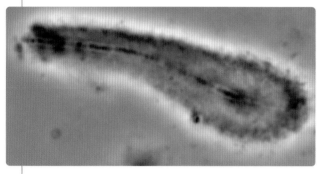

*Top* These spherical-shaped iron bacteria are like *Gallionella*. They attach to underwater surfaces and secrete mucus that accumulates metal salts, especially iron, as the cells grow.

*Above* This enlarged image shows details of the mucus shell that surrounds *Gallionella* cells.

## Photosynthetic Bacteria

There are many types of photosynthetic bacteria, including cyanobacteria, green nonsulfur bacteria, and purple sulfur bacteria, which we'll look at below.

## Cyanobacteria

Cyanobacteria are a large and diverse group of photosynthetic bacteria. They are often the dominant photosynthetic organisms in many microbial communities found in Yellowstone. Cyanobacteria grow in many extreme environments such as desert soils, saline lakes, warm seawater, and hot springs. Many other species occur in freshwater lakes, in marine habitats, and even inside porous rocks. In Yellowstone, cyanobacteria form or contribute to extensive, thick mats and thrive in water with temperature as high as 74°C (165.2°F), the upper limit for photosynthesis.

Cyanobacterial cells possess distinctive colors due to pigments used for photosynthesis or for protection from high levels of sunlight. Cyanobacteria have only one form of chlorophyll, chlorophyll *a*. They also contain other accessory pigments that help in photosynthesis. One of these is a distinctive cyan-colored pigment that, in combination with chlorophyll, gives some species of cyanobacteria a blue-green color. Cyanobacteria were once referred to as the "blue-green algae" because of this distinctive color. Other pigments produce a deep red or brown color in some species.

Different species have various shapes such as rods, **cocci** (spheres), spirals, or filaments. Colonies of cyanobacteria can grow as thin, slimy sheets over surfaces or long filaments that intertwine to form jellylike mats. The mats can trap grit, sinter, and other debris in layered deposits. Filamentous cyanobacteria often grow with other photosynthetic bacteria in structures called stromatolites.

Many cyanobacteria can move by gliding to find habitats with ideal conditions for growth. They respond to changing levels of sunlight and oxygen and to variations in temperature.

Cyanobacteria are also important because some species can "fix" nitrogen. Nitrogen is a key component of proteins and genetic material, but only a few microbes have the ability to take up nitrogen from the environment and incorporate it into biological compounds. All other organisms depend upon atmospheric nitrogen fixed by bacteria into organic forms of nitrogen that can be used for cell growth.

Cyanobacteria have played an important role in the history of the Earth. During photosynthesis these organisms produce oxygen as a by-product. It was due to the emergence of the cyanobacteria about two billion years ago that the Earth's atmosphere changed from one without oxygen to one with oxygen. Cyanobacterial production of oxygen sometimes can be observed in hot springs. Bubbles can lift a microbial mat up into vertical structures called stromatolites. If left undisturbed, some stromatolites can develop into substantial structures visible to the human eye.

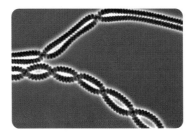

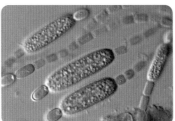

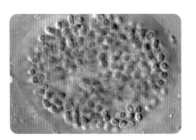

*Top to bottom*

*Spirulina* has a distinctive corkscrew shape.

*Cylindrospermum* is composed of many cells joined end to end. In this species, the larger cells perform nitrogen fixation.

This small clump of the cyanobacterium *Microcystis* is surrounded by mucus. Some cyanobacteria that secrete mucus produce toxins that affect animals and humans who drink or bathe in the contaminated water.

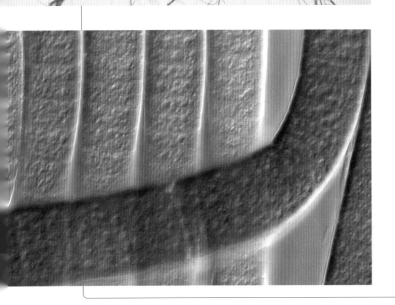

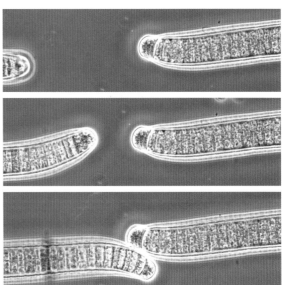

*Left top* The long, intertwined filaments of cyanobacteria grow together to form a microbial mat.

*Left bottom* Some cyanobacteria have filaments that are made up of many disc-shaped cells joined end to end. The colors of the filaments can vary even when they are composed of cells of the same species. This can cause variations in the colors of the mats.

*Below* *Oscillatoria* is able to move by gliding over surfaces. These three images were taken over a period of about two minutes and show two filaments moving toward each other.

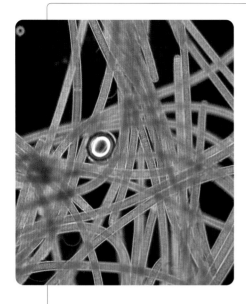 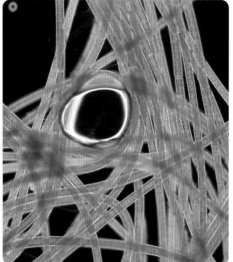 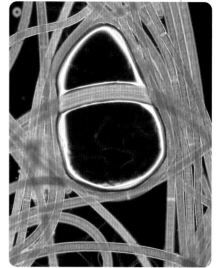

*Top* These three images taken over a period of five minutes show the formation of an oxygen bubble produced as a result of photosynthesis by the cyanobacteria.

*Left* These vertical columns in a cyanobacterial mat were pushed upward by bubbles of oxygen trapped within the mat. The columns are living stromatolites, composed of layers of bacteria and trapped sand. Over time stromatolites can harden and form rock.

*Below* A highly magnified *Synechococcus* cell.

*Bottom* These rod-shaped *Synechococcus* cells contain bright green photosynthetic pigments.

## ● Synechococcus

*Synechococcus* is a common and abundant type of rod-shaped cyanobacterium with light-colored regions at the ends of the cell. In Yellowstone, it is most noticeable as a green component in extensive orange mats such as those found in La Duke Spring just north of the park or Octopus Spring in the Lower Geyser Basin.

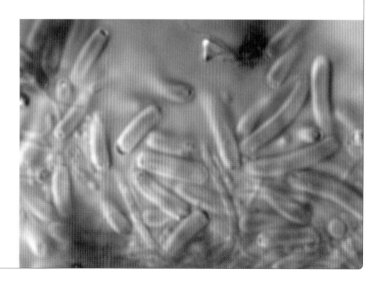

## Green Nonsulfur Bacteria

Yellowstone is home to a distinctive type of green nonsulfur bacteria called *Heliothrix*. Even though *Heliothrix* is classified as a green nonsulfur bacterium, it contains a distinctive bacterial chlorophyll that gives it a red color. When large numbers of *Heliothrix* grow in alkaline springs at temperatures from 35°C to 60°C (95°F to 140°F), they form red-layered mats. These red layers can be seen at Witch's Pond near Heart Lake or at Hillside Springs in the Old Faithful area. *Heliothrix* and other green nonsulfur bacteria, including a species of *Chloroflexus,* are filamentous, photosynthetic microbes that grow in layers of mats where there is little oxygen. *Chloroflexus* is important because it and its relatives are the primary component of many of the extensive orange mats found in thermal features. These photosynthetic bacteria may be similar to ancient photosynthetic bacteria like those preserved in fossilized stromatolites.

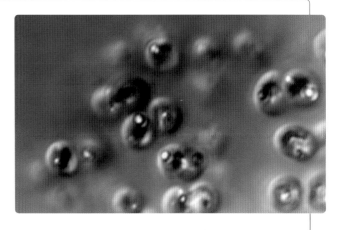

## Purple Sulfur Bacteria

The purple (or red) sulfur bacteria capture sunlight for energy in cell metabolism. They are photosynthetic organisms, but unlike the cyanobacteria and algae and plants, they do not produce oxygen as a by-product. Purple sulfur bacteria use hydrogen sulfide to derive energy and produce pure sulfur as a by-product. The sulfur is stored inside the bacterial cells. They capture energy with pigments that give these microbes red, brown, or purple colors. Purple sulfur bacteria are found in habitats that receive sunlight but have little oxygen, such as the bottoms of shallow streams and ponds. These habitats are characterized by the intense smell of rotten eggs from hydrogen sulfide gas. The purple sulfur bacteria, with their ability to undergo photosynthesis without oxygen, appeared earlier in Earth's history than the cyanobacteria.

*Below* A small piece of the orange mat from Lakeside Spring in West Thumb Geyser Basin contains several species of bacteria. *Chloroflexus* grows in long, thin filaments.

*Above* Purple sulfur bacteria are important for sulfur cycling in aquatic ecosystems like Swan Lake. The bacteria thrive in mud that has very little oxygen. The bright spots are sulfur granules inside the cells.

# Other Bacteria

Bacteria, like other organisms, must be able to react quickly to environmental changes, such as variations in temperature or sunlight. Many bacteria can glide (move by creeping slowly) across a surface. Others swim by spinning stiff, propellerlike hairs.

Bacteria come in many shapes and sizes. Some are spherical. Others are rod or sausage shaped. Often bacteria form aggregates, sometimes bound together within a gelatinous material. Other

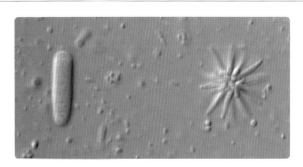

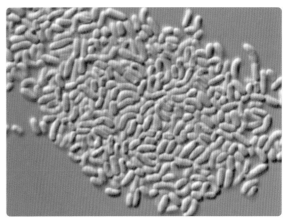

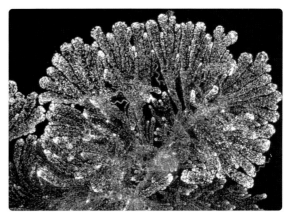

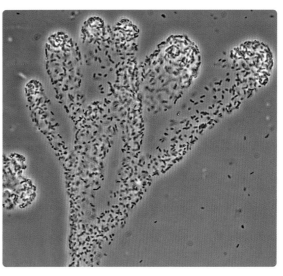

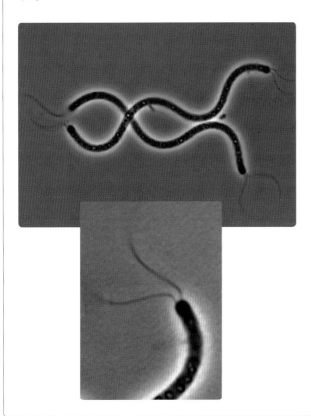

*Right, top to bottom*

Numerous microbes live in a bison's stomach and aid in digestion. Shown here are a star-shaped clump of cells, a thick rod-shaped cell, and many small round bacteria.

Thermophilic, pink-colored bacteria are among the many species that live in hot springs.

These bacteria are growing in a jellylike colony. This form of slimy growth is common in environments with lots of organic matter such as sewer sludge.

The rod-shaped bacterial cells in this image are held together by delicate mucus.

*Below* *Spirillum* cells have stiff, corkscrew-like flagella that move the cells through water.

*Below inset* This image shows the two flagella attached to a *Spirillum* cell.

types attach firmly end to end to create long filaments. Bacteria may be transparent or colored. Cyanobacteria and purple sulfur bacteria have pigments that capture sunlight. Other bacteria contain pigments with no known function.

Many bacteria use the abundant sulfur found in Yellowstone as an energy source.

# Eukarya

## ● Algae

Algae abound in freshwater and saltwater and in soils. Some live as **symbionts** (organisms that live with another organism) in microbes such as paramecia or fungi, or in plants. Algae, like the higher plants, carry out photosynthesis, producing starches and sugars for growth and releasing oxygen as a by-product.

Kelps and other seaweeds are algae that occur in marine environments and can grow to more than 50 meters in length. Many kinds of algae, however, are microscopic single cells. Single cells can grow together to form visible colonies or mats. Algae have rigid cell walls that give them distinct shapes. Some algae live in soil or in porous rocks, but most live in aquatic habitats. They contribute greatly to the food web as primary food producers for other organisms.

Microscopic algae can live individually, in spherical colonies, or stacked end to end to form long, visible filaments. Chlorophyll and other pigments give algae distinctive green, brown, or red colors. Aggregates of microbial algae can be seen as bright green scums in ponds, on rocks, and on trees, or as a green or orange slime on mud.

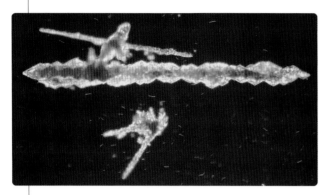

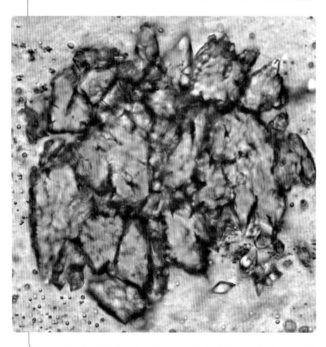

*Top to bottom*

These curved bacteria are covered by sulfur.

This image was taken with a powerful electron microscope. The small sample of a bacterial mat collected in Norris Geyser Basin contains diamond-shaped crystals of sulfur. Spaghettilike bacteria surround the sulfur crystals. *Image provided by W. Inskeep and color enhanced by D. Patterson.*

Crystals of sulfur examined under a microscope often reflect intense colors.

# Green Algae

The green algae are a large group of microscopic algae that live in freshwater habitats or in soils. Green algae have thick cell walls and a distinctive bright green or yellow green color.

Filamentous green algae in Yellowstone form conspicuous mats along the edges of many nonthermal lakes and streams. Each algal cell typically has a central nucleus and one or more chloroplasts. In *Spirogyra*, the chloroplast looks like a thick, curly ribbon just inside the cell wall. Others like *Ulothrix* and *Klebsormidium* have pancakelike chloroplasts pressed against the inside of the cell wall.

*Background*
A green algal mat thrives along the shoreline of a pond near Firehole Lake Drive.

*Above inset*
Bright green scum formed by algae often grow in the shallow water at the edges of ponds and lakes.

*Top left*
*Klebsormidium*, a filamentous green alga, grows in loose clumps.

*Middle left*
Filamentous green algae contain bright green chlorophylls that trap sunlight for photosynthesis. The green structures in the center of the *Klebsormidium* cells are the chloroplasts where the pigments are located.

*Bottom left*
In *Spirogyra*, the chloroplasts spiral around the interior of the cell. The nucleus is slightly to the right of the center. This alga was collected from the Gardner River near Sheepeater Cliffs.

*Background* This leathery dark maroon mat growing at Nymph Lake is mostly formed by a green alga called *Zygogonium*.

*Inset top* Insect larvae feed on the *Zygogonium* mats.

*Inset bottom* Magnified *Zygogonium* filaments show the cell walls that separate individual cells and the green chloroplasts where photosynthesis takes place. These chloroplasts are surrounded by red vacuoles that often give clumps of cells a maroon color.

*Zygogonium*, classified with the green algae, lives in and around acidic streams and lakes in places like Nymph Lake and the Norris Geyser Basin. Individual *Zygogonium* cells contain a dark maroon-red pigment that protects the alga from too much sunlight. The cells grow end to end in long strands that often intertwine in thick leathery mats. *Zygogonium* survives very acidic conditions and cannot grow above 35°C (95°F).

Many green algae are spherical. *Chlorella* is an example of this form. Some *Chlorella* species are

*Right* These deep-green *Chlorella* cells grow in acidic water like Nymph Creek and Lemonade Creek but do not grow at temperatures higher than 35°C (95°F). Several *Cyanidium* cells, with their blue-green pigments, are also visible.

*Far right top* *Chlamydomonas* is a common green alga that thrives in freshwater habitats like Swan Lake. Each cell has two flagella and a bright red eyespot used to sense sunlight.

*Far right bottom* These two *Chlamydomonas* cells are undergoing conjugation, a form of sexual reproduction in protozoa. The cells exchange genetic material and produce a new generation of cells with a different genetic makeup.

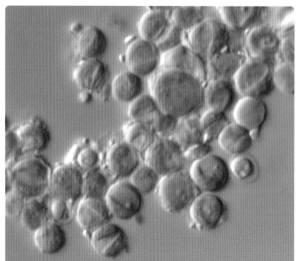

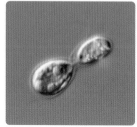

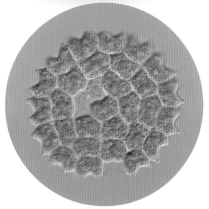

symbionts inside protozoa, such as the ciliate *Paramecium*. Other species live in animals like the green hydra. *Chlorella* cells do not actively move; they are carried by currents through the water.

*Chlamydomonas* is a spherical or elongate-shaped green alga. *Chlamydomonas* cells have long filaments called flagella that vibrate very quickly and move the algae through the water. *Chlamydomonas* reproduces by dividing into two new cells. Occasionally, cells join end to end to exchange genetic material.

**Desmids** are microscopic green algae that often look like two mirror-image cells joined together. The cells share a nucleus and a chloroplast. Desmids live in clean and slightly acidic water, conditions often found in pristine alpine lakes. Some desmids, such as *Pediastrum*, have a more complex star shape.

*Cladophora* cells branch in treelike shapes. *Cladophora* is common in fast-flowing rivers such as the Gardner or Yellowstone.

*Top* Desmids are single-celled green algae that live in freshwater. They often have beautiful, symmetrical shapes like this *Cosmarium* found in South Twin Lake.

*Center* There are many types of desmids, such as this distinctive smile-shaped *Closterium* collected at Obsidian Creek.

*Above* This star-shaped desmid is a colony composed of individual *Pediastrum* cells.

*Below right* This microscopic image shows the cell walls of individual *Cladophora* cells.

*Below left* The green alga *Cladophora* is common in freshwater streams. The cells join end to end and form mats.

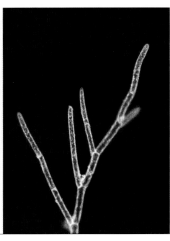

## Diatoms

Diatoms are one of the most diverse, abundant, and important of all the groups of microscopic algae. So far more than 100,000 species have been identified. They occur in profusion in freshwater, marine, and soil habitats.

Diatoms are some of the principal organisms carrying out photosynthesis in the oceans. Current estimates are that diatoms may account for as much as 40 percent of the carbon fixation (the combining of carbon dioxide and water during photosynthesis to produce food) that occurs in the world's oceans, releasing oxygen as a by-product, and contributing significantly to the amount of oxygen in the Earth's atmosphere.

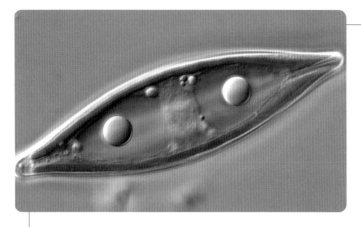

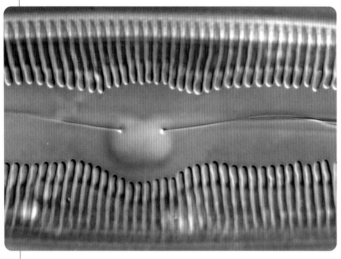

Diatoms have two basic shapes. The centric diatoms are cylindrical and shaped like a round pillbox, and the rod-shaped **pennate** diatoms look like small boats. Diatoms contain one or more golden or brown chloroplasts. The cells are enclosed in a glassy shell (or case) called the frustule. The frustule is perforated by tiny holes, usually in regular geometric patterns. These patterns allow specialists to identify the various species; the patterns also give diatoms an exquisite beauty. In the center of each face of the frustule of a pennate diatom is a delicate groove, called the **raphe.** Mucus is pushed out through this groove, allowing the diatom to glide back and forth in the direction of the groove.

The glass frustules do not break down quickly after a diatom dies. They settle on the bottoms of oceans, lakes, and ponds. Over time, the accumulation of frustules can be enormous, creating economically valuable deposits that can be mined and used for

*Top* This beautiful diatom is enclosed in a glass shell known as a frustule. The chloroplasts are the brown structures that are pressed up against the inside of the cell wall. The nucleus lies in the center of the cell. The bright globules are drops of oil, stored as a future energy source.

*Above* This image shows the intricate frustule of the diatom *Pinnularia*. Frustules have ridges, grooves, and pores that are useful for identification of different species. The two grooves in the center of the cell form the raphe, which is used for propelling the diatom.

*Below* Diatom frustules are brittle and delicate. This *Pinnularia* frustule shattered when it was flattened between the microscope slide and cover slip.

*Right* This diatom has an intricate frustule.

*Inset bottom* These two cells are the same species of diatom. The central nucleus and chloroplasts are visible in the lower cell. The center groove, or raphe, is visible along the surface of the upper cell.

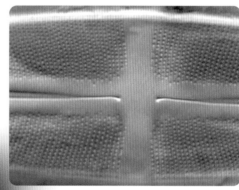

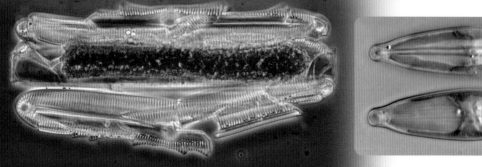

**Seen *and* Unseen**

abrasives in industry—as an ingredient in toothpaste, for instance. Many ocean diatoms sink to the sea floor as "marine snow." The accumulation of debris forms a greenish scum that provides food for other organisms that live on the ocean floor. Yellowstone Lake spires, pinnaclelike formations recently discovered on the floor of Yellowstone Lake, are composed largely of diatom shells.

## ● Cyanidium

*Cyanidium caldarium* is the most acid-tolerant alga known and can grow in the laboratory at a pH as low as 0.5. In Yellowstone, algae such as this frequently form mats in hot springs and streams and along steam-covered rock surfaces at pH 3 and temperatures around 45°C–50°C (113°F–122°F). *Cyanidium caldarium* and its close relatives form green mats in streams such as Nymph Creek. They contain a bright green photosynthetic pigment, **phycocyanin,** which gives them a distinctive bright green color. Despite this color, *Cyanidium* species belong to the group Rhodophyta, the red algae, and are related to the red marine seaweeds.

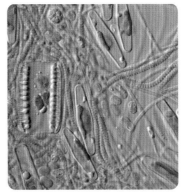

*Left* A piece of a microbial mat collected from Orange Spring Mound contains intertwined cyanobacteria and two kinds of diatoms, *Achnanthes* and *Surirella*.

*Below* Each cell of this centric diatom species collected from the Yellowstone River is a small cylinder that joins with another to form a long chain of cells. These cells cannot move. The green structures inside the cells are the chloroplasts.

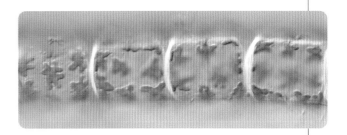

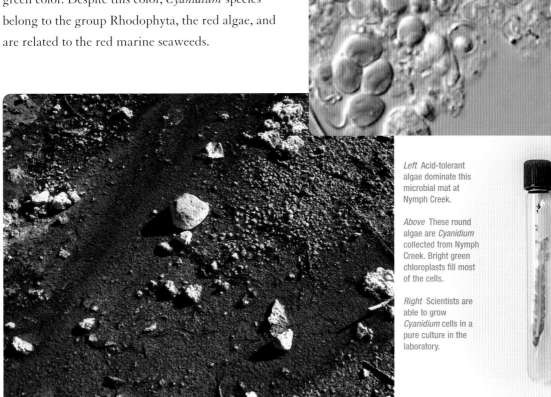

*Left* Acid-tolerant algae dominate this microbial mat at Nymph Creek.

*Above* These round algae are *Cyanidium* collected from Nymph Creek. Bright green chloroplasts fill most of the cells.

*Right* Scientists are able to grow *Cyanidium* cells in a pure culture in the laboratory.

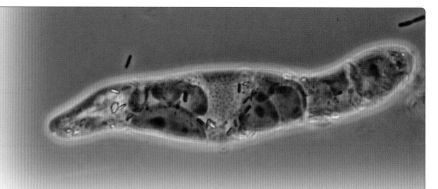

This *Euglena* cell collected from Beaver Lake emphasizes the disk-shaped green chloroplasts. The front of the cell is to the left. The light area is called the reservoir. Adjacent to this region is the red eyespot that helps to control the direction the cell moves. The granular region in the center of the cell is the nucleus.

## Euglenids

Euglenids are classified as algae by botanists because many species have chloroplasts and carry out photosynthesis. There are about one thousand different described species of euglenids. Most species of *Euglena* have one or more bright green chloroplasts and have two flagella; often only one flagellum is found at the front of the cell. This flagellum beats with a whiplike motion that propels the cell forward. Other species of *Euglena* stay attached to surfaces. *Euglena* have a bright red eyespot near the front of the cell that is used to detect sunlight and stimulate the flagellum to move the cell toward favorable light conditions.

*Euglena mutabilis* is common in acidic environments in Yellowstone. It contains four to eight disk-shaped, bright green chloroplasts and a bright red eyespot. This *Euglena* moves by writhing and squirming through mud or microbial mats.

*Peranema* is a euglenid that looks like an amoeba, but it has a single flagellum at the front of the cell that barely moves except at its tip. *Peranema* has no chloroplasts for photosynthesis. It lives as a predator, attacking and eating other protozoa. *Petalomonas* is another common type of euglenid found in similar habitats.

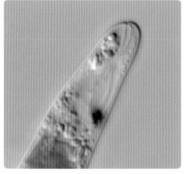

*Top left* This euglenid cell, collected at Beaver Lake, has a clear region at the front that contains a short flagellum near the bright red eyespot. The flagellum does not extend out of the cell.

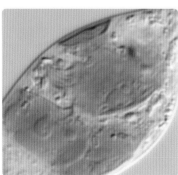

*Bottom left* The bright green, disk-shaped chloroplasts are clearly visible in this *Euglena* collected from Beaver Lake.

*Left Phacus*, found in Obsidian Creek, is a flat, stiff type of euglenid with many small disk-shaped chloroplasts and a red eyespot. The clear area in the cell is the nucleus.

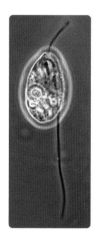

*Right* This euglenid, called *Anisonema*, was collected from Obsidian Creek. It has two flagella. One, at the front of the cell, sweeps from side to side. The second trails behind the cell.

*Bottom left Peranema* has a thick flagellum. It has no chloroplasts for photosynthesis. Instead, it is a predator that eats other microbes.

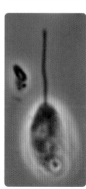

*Bottom right Petalomonas* is another type of euglenid collected from Obsidian Creek.

## Protozoa

Protozoa are a large group (more than 65,000 known species) of typically nonphotosynthetic microbes that do not have rigid cell walls. Most protozoa contain no pigments and are colorless. Nearly all are able to move to catch food and are classified into groups according to the type of locomotion they use. Most protozoa do not have chloroplasts and ingest food made by other organisms. Some protozoa cause illnesses such as sleeping sickness or malaria. Species found in natural habitats are called "free living" and are grouped into several types based upon how they move and capture food.

Amoebae form **pseudopodia,** footlike extensions that they use for pulling along surfaces and for surrounding and engulfing food such as bacteria or algae. Ciliates swim by rapidly beating hairlike structures called cilia, found in rows along the surface of the cells. **Flagellates** move by vibrating one or more thin appendages called flagella. Some protozoa appear to glide by means of undulating ridges along the cell surface.

## Amoebae

Amoebae extend and retract temporary pseudopodia (false feet) from the surface of the cell to move through the environment or to capture food. Most amoebae feed on smaller protozoa, bacteria, and particulate material in streambeds or on slimy surfaces.

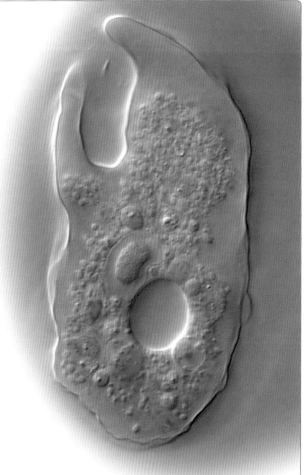

*Above* Thecamoeba is a stiff type of amoeba that extends temporary threadlike filaments called pseudopodia (false feet) from the cell. The pseudopodia are used to trap food or for movement. The large bubblelike structure is a contractile vacuole that helps to control the water content in the cell.

*Below left* Dactylamoeba has many tapering pseudopodia. The hyaline cap, a clear region at the lower right front of the cell, is the place where new pseudopodia emerge.

*Below center* This red amoeba eats thermophilic algae in places like Nymph Creek. Partially digested algae are inside the cell.

*Below right* Nuclearia has many thin pseudopodia that it uses to move or to trap food such as algae.

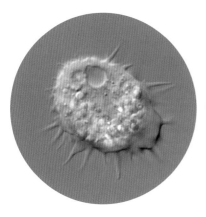
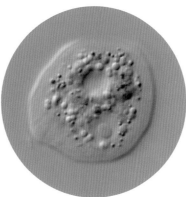
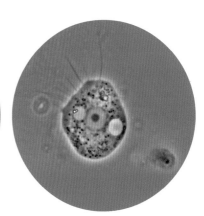

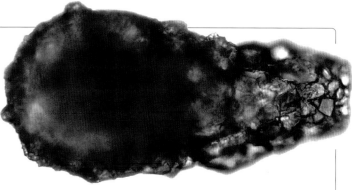

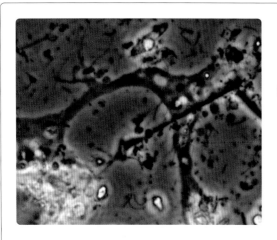

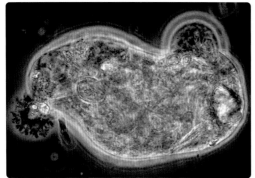

Cell shape constantly changes as the amoebae move. Typical amoebae have a single nucleus and a contractile **vacuole** (a bubblelike compartment in a cell). The contractile vacuole helps to remove excess water from inside the cell.

Different types of amoebae are identified by the type and number of pseudopodia. Pseudopodia vary from being very broad and rounded, to threadlike (filose) or tapering.

Amoebae can be classified based upon any coverings or shells that surround the cell. Coverings may provide protection for the amoebae. *Cochliopodium* is covered in a layer of very small scales. *Pompholyxophrys* has a coating of small grit.

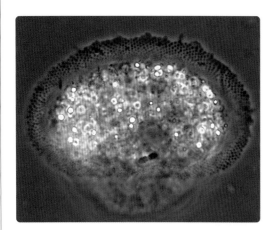

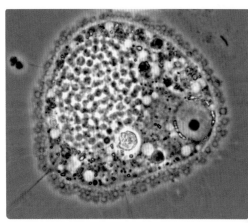

*Left column, top to bottom*

This rare type of amoeba was collected from a mineral deposit in a hot (45°C or 113°F) region of Nymph Creek. Sometimes, as in this image, many amoebae fuse together in a large group. Often these groups can be many millimeters in size.

*Pelomyxa* is an unusual amoeba because it lives in habitats with little oxygen, such as mud on the bottom of a lake or pond. The cell contains ingested algae and sand grains.

*Cochliopodium* is covered with delicate scales that look like small black dots around the edge of the cell.

*Pompholyxophrys* is an amoeba that is covered with delicate, glasslike spheres.

*Top right* This amoeba, *Difflugia*, makes a shell using small particles of grit.

*Below* This image shows an orange center hole at one end of a *Difflugia* cell where four pseudopodia have emerged.

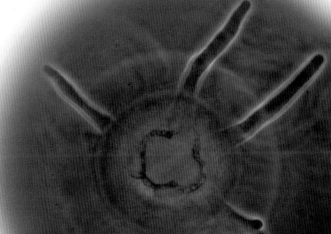

**Seen** *and* **Unseen**

Some amoebae, called "testate" or "shelled" amoebae, have a more substantial shell or case, usually with a single opening for the pseudopodia. Amoebae without shells are referred to as "naked."

The easiest amoebae to identify under a microscope are the heliozoa. They were described as "sun-animalcules" by early biologists because they have a nearly spherical body with stiff radiating pseudopodia that look like the rays of the sun. Heliozoa are predators that barely move. Instead they trap microbes that bump into them when a sticky substance is released from the radiating pseudopodia. Once the prey is trapped, more pseudopodia extend outward, wrap around the prey, and enclose it in a vacuole where digestion takes place.

The vahlkampfiid amoebae occur in freshwater and soil habitats around the world. They have the ability to change from the amoeba cell form to a flagellate form. Sometimes they form **cysts**—a dormant stage that protects the amoeba from harsh conditions—for protection from drying out. A few species are thermophilic and able to thrive in hot springs and man-made water systems such as water heaters, heated swimming pools, and spas. Given the right conditions, some of these amoebae can infect humans and other mammals, causing eye infections, diarrhea, or serious brain infections.

One thermophilic species, *Naegleria fowleri*, causes a rare but fatal disease, primary meningoenchephalitis. Infection occurs after ingesting contaminated water in the nose, usually when swimming or diving. The amoebae travel along the olfactory nerves and enter the brain, resulting in severe symptoms and death ten to fourteen days after exposure. These amoebae have been identified in some thermal waters in Yellowstone.

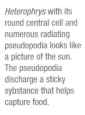

*Heterophrys* with its round central cell and numerous radiating pseudopodia looks like a picture of the sun. The pseudopodia discharge a sticky sybstance that helps capture food.

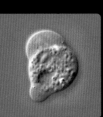

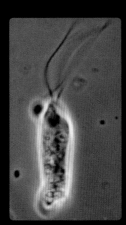

*Top left* This type of amoeba, called a vahlkampfiid, was collected from Nymph Lake. The clear region at the upper left of the cell is where a pseudopodium forms.

*Left* Some vahlkampfiid amoebae can change into a different form with flagella.

*Right* Vahlkampfiid amoebae in the genera *Naegleria* and *Acanthamoeba* are able to cause serious infections of the eye and sometimes fatal brain infections in humans and other animals. The amoebae live in heated water at temperatures that humans prefer for swimming and soaking. This sign posted at Boiling River warns the public about the health risks.

> Thermally-influenced waters may contain bacteria known to cause serious skin rashes, infections, and/or amoebic meningitis which can be quickly fatal. Ingesting water through your mouth and nose is particularly dangerous. Keep your head above water; if you have any signs or symptoms of irritation or disease, seek medical help immediately.
>
> **SWIM AT YOUR OWN RISK**

## Flagellates

Flagellates move by actively beating one or more long, thin flagella. Most are small compared to other protozoa, but they are extraordinarily abundant. A typical teaspoonful of seawater contains several thousand flagellates. Flagellates are important in the microbial food web and help to keep water healthy by consuming bacteria that would otherwise grow in large numbers, use up oxygen, and cause water to become putrid.

Some flagellates cause diseases in humans. Giardiasis is a gastrointestinal disease caused by a flagellated protozoan parasite, *Giardia lamblia*. The parasite is present in lakes, ponds, and streams and is transmitted by drinking contaminated water.

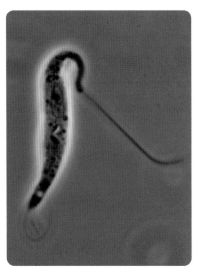
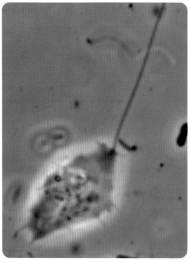
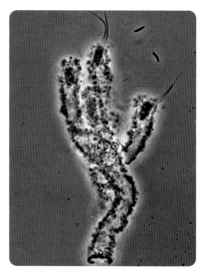

*Above left* This microbe, collected from Swan Lake, was previously unknown. There is still much to be learned from exploring the natural world.

*Above center* This microbe, *Mastigamoeba*, looks like an amoeba, but it belongs to the flagellates. It has one long flagellum at the front of the cell.

*Above right* *Siphomonas* is a flagellate that lives in colonies made from tubes of mucus. Each *Siphomonas* cell has two flagella.

*Below left* This small flat flagellate with two flagella is called *Goniomonas*.

*Below center* *Bodo* is a common flagellate found in many habitats. The cell is shaped like a kidney bean. There are two flagella. One is stretched along the right side of the body in this image. The other flagellum is longer than the cell and is used to attach *Bodo* to a surface.

*Below right* *Rhynchomonas* is called the nosey flagellate. Its mouth is shaped like a nose. This microbe eats bacteria.

## Ciliates

Ciliates are relatively big for microbes. Most are more than 50 microns in size and some, such as *Spirostomum*, grow to a millimeter or more in length. Ciliates are fast-moving swimmers with a variety of shapes and colors. As a result, they are eye-catching when viewed under a microscope.

Ciliates move and feed by beating short hair-like filaments, called cilia, arranged in rows along the cell's surface. Each cilium beats slightly out of phase from its neighbor, creating a coordinated wavelike movement of water across the cell's surface. In a few types of ciliates, the cilia are bundled together in clusters called **cirri.** The cirri act like legs pressing against solid surfaces, allowing the ciliate to "walk" over debris and mud rather than swim.

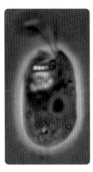

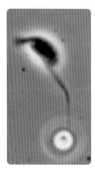

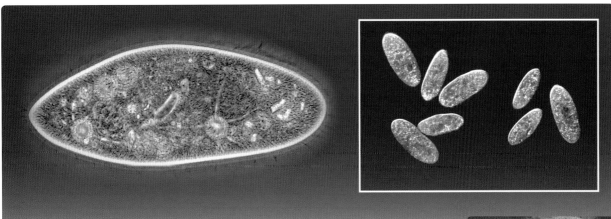

Ciliates are classified into three groups based upon the types of food they consume. Some are hunters and feed on moving prey such as algae and other protozoa. They can inject the prey with toxins or can catch the prey with specialized hooklike structures. Others eat relatively large stationary particles such as detritus and algae. The third group contains filter feeders that catch suspended food particles, including bacteria. These ciliates have special groups of cilia that sweep food into a mouthlike region called the **cytostome.**

Although most ciliates feed on other organisms, some contain large numbers of green algae. This symbiotic relationship allows both of the partner organisms to benefit. The algae get nutrients and protection, and the ciliates obtain food and oxygen as a result of photosynthesis.

*Paramecia* are described in many basic biology textbooks. Because they are easy to grow in the laboratory, they are studied in biology classrooms worldwide. All species of *Paramecia* have thousands of cilia covering the cell's surface. Underneath these cilia are large numbers of short, rod-shaped organelles

*Clockwise from upper left*

This *Paramecium*, covered with hundreds of cilia, was flattened slightly to make the cell components more visible under the microscope.

*Paramecia* are well known to scientists and biology students because they are common in many habitats and are easy to grow in a laboratory.

Cilia used for the movement in *Brachonella* are arranged in rows, called kineties, around the cell. Other cilia at the bottom of the cell are used for feeding.

Many rows of cilia cover *Brachonella*.

Some cilia (at the bottom of the picture) of *Stylonychia* are clustered together and used like legs. Another type of cilium at the front of the cell is used to gather food.

This unusual ciliate, called *Aspidisca*, also has cilia that form legs.

The striking cilia surrounding *Halteria* are used for feeding.

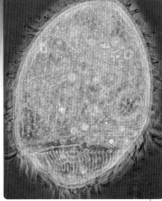
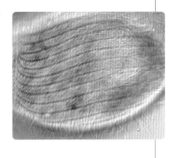
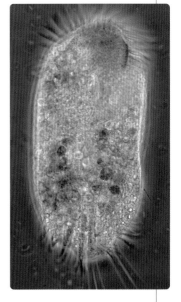
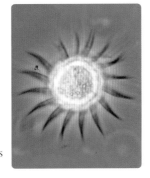
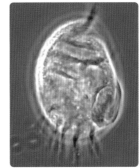

*The Microbes*

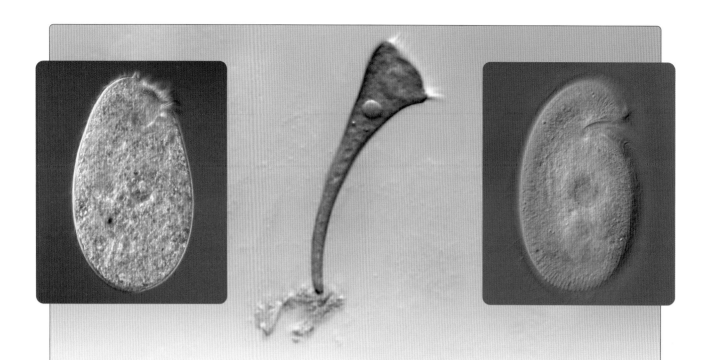

(membrane-enclosed structures in eukaryotic cells) called **trichocysts** or **extrusomes**. These can explode and shoot outwards as stiff hairs. If a *Paramecium* is under attack from a predator or is in contact with harsh chemicals, it uses these organelles to rapidly propel itself away from the danger. *Paramecia* usually eat bacteria by filtering water with cilia located around the mouth. The food passes into the mouth and is packaged in membranes to form food vacuoles where digestion takes place.

Eukaryotic cells are characterized in part by organelles inside the cell. In particular, eukaryotes have a nucleus that contains the cell's genetic material. Most eukaryotic cells have a single nucleus. Some, especially some protozoa, may have many similar nuclei in one cell. The ciliates are unusual because they have two kinds of nuclei. One of them, the larger macronucleus, is used to generate protein and carry out the functions of the cell. The other nucleus is smaller. It contains genetic material

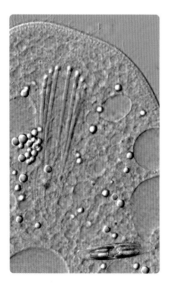

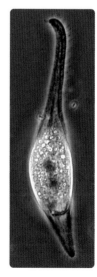

*Top left*
*Climacostomum* has special feeding cilia at the front of the cell (top of image). There are many bright green symbiotic algae in this cell.

*Top center*
Remarkable *Stentor* cells pulse by contracting or stretching out, sometimes to half a millimeter in length. Cilia at the top of the cell help collect food.

*Top right*
*Plagiopyla* lives in habitats with little or no oxygen. This ciliate prefers to eat sulfur bacteria such as *Beggiatoa*.

*Left*
This ciliate eats diatoms. The cell has a mouth with tough jaws that grab diatoms and force them into the cell. One diatom is in the lower right of the cell.

*Right*
*Spathidium* preys on other ciliates by shooting special organelles from the mouth region.

*Far right*
*Litonotus*, collected from South Twin Lake, is a predatory ciliate that shoots out special organelles to kill prey. The two dark structures in the middle of the cell are large nuclei.

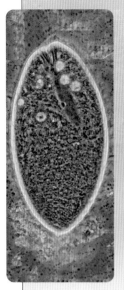 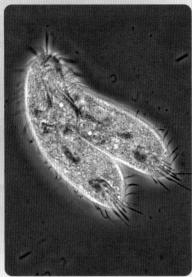

# Fungi

Fungi are common in diverse habitats, found in aquatic environments, in soils, on dead plant matter, or as parasites of plants and animals. Fungi are eukaryotic, nonphotosynthetic organisms divided into three major groups and include the molds, the yeasts, and the mushrooms. Fungi often produce microscopic spores that are easily spread in the environment. Many are **saprophytes**—that is, they obtain nutrients by decomposing dead and dying plants and animals.

The fungus *Penicillium* was the first source of the antibiotic penicillin, a potent disease-fighting drug. Many fungi are used commercially for the manufacture of foods such as blue cheese, soy sauce,

that is used during sexual reproduction. Sexual reproduction usually involves two cells fusing near the mouth and then exchanging a copy of the genes from the smaller nucleus.

Contractile vacuoles are used to control the water content of the cell by picking up excess water from the cytoplasm and delivering it to the vacuoles. The vacuoles then periodically discharge their contents out of the cell.

*Top left*
This isotrich ciliate was obtained from a bison stomach. The angled gray structure near the front of the cell (top of picture) is the macronucleus. The small dark structure pressed in the left side of the macronucleus is the small nucleus.

*Top right*
These ciliates, collected from South Twin Lake, are undergoing conjugation, a sexual exchange of genetic material.

*Bottom left*
*Dactylaria* is a fungus that tolerates hot environmental conditions. This fungus was isolated from soils at Amphitheater Springs. The thin hyphae grow spores (the oblong-shaped cells).

*Bottom right*
Many fungi have long branching cells called hyphae.

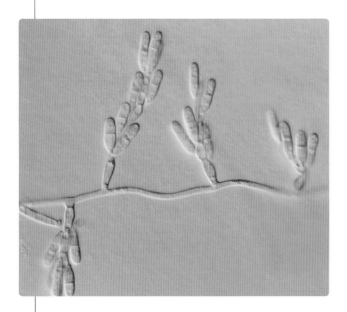 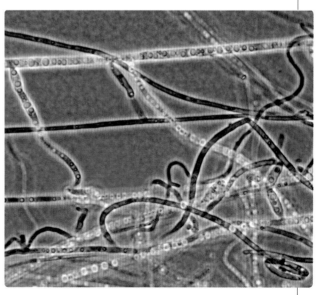

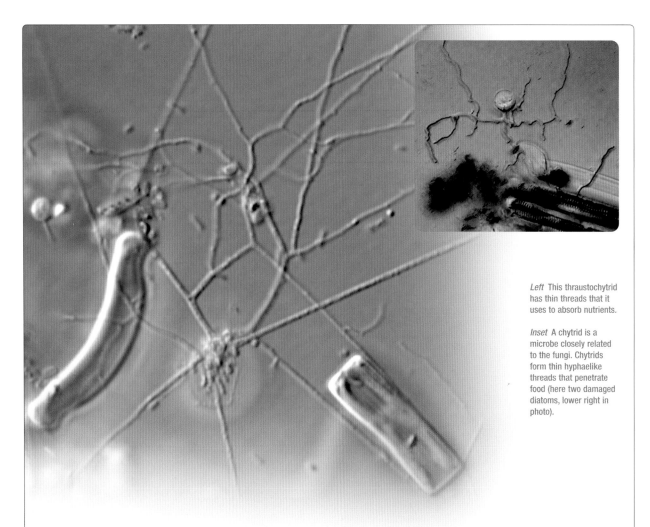

*Left* This thraustochytrid has thin threads that it uses to absorb nutrients.

*Inset* A chytrid is a microbe closely related to the fungi. Chytrids form thin hyphaelike threads that penetrate food (here two damaged diatoms, lower right in photo).

bread, beer, and wine. A number of fungi are adapted to extreme environments like the solfataras found in Yellowstone.

Molds form filamentous cells with rigid walls in long structures called **hyphae.** Hyphae grow together to form a visible, fuzzy clump called a **mycelium.** The mycelium sometimes produces brightly colored spores that may be black, brown, blue, green, red, or yellow due to pigments. Molds are common in soils, in compost heaps, on cheeses, and on stale bread.

Yeasts are microscopic, single-celled fungi that grow in environments where sugars for growth are present. They are found in soils and sometimes cause infections in humans. Yeasts known as *Saccharomyces* are important commercially for bread baking and brewing.

Mushrooms are spore-bearing fruiting bodies produced above the ground by some fungi. For most of their life cycle, the fungi live as microscopic mycelia underground or in plants. When conditions are favorable, the mycelia form a mushroom above ground that produces spores. The spores are released into the environment to colonize new sites.

Fungi may have evolved from a group of protists called **chytrids.** Chytrids eat algae or other dead organisms in aqueous environments. They settle on the surface of a food source and grow fine tubes (**rhizoids**) that penetrate into the food in order to absorb nutrients. The chytrid grows and eventually divides into small flagellated cells that seek out new sources of food.

# The Relationships

Free-ranging bison herds graze in the meadows of Yellowstone National Park.

John Donne said: "No man is an island, entire of itself; every man is a piece of the continent, a part of the main." The same can be said of the tiny world of microbes. They interact in ecosystems with each other and with plants and animals. These associations can affect environments in many ways.

Microbes provide essential nutrients for higher animals and plants. All organisms require nitrogen, but only a few species of bacteria have the ability to convert atmospheric nitrogen into compounds that other organisms can use. In legumes such as beans, peas, and alfalfa, symbiotic bacteria fix nitrogen in the root nodules, producing **nitrates** (chemicals required for making proteins).

Microbes play vital roles in cycling nutrients in freshwater and marine ecosystems. Various microbes act as primary producers at the bottom of food chains. These organisms capture sunlight or chemical energy to change or "fix" carbon dioxide and inorganic compounds into organic matter that is used by other organisms for nutrition and energy. In freshwater and marine ecosystems, photosynthetic microbes are the major primary producers.

When aquatic organisms die, much of the organic matter that makes up their bodies sinks to the bottom of a lake or stream where it is decomposed by microbes, depleting oxygen and creating a layer on the bottom where anaerobic bacteria are able to flourish. These anaerobic bacteria act as decomposers and return nutrients and inorganic compounds back to the environment to be used again by primary producers. Microbes, therefore, are important as consumers and producers. These

interactions by microbes create the microbial food web, the most significant component of any ecosystem.

The interactions of organisms and the roles they play in ecosystems is not a simple story. There are many ways organisms affect each other and their communities. Competition for food, space, light, favorable temperatures, and avoidance of disease are some factors that are just as important for microbes as for plants and animals. Organisms have evolved with complex relationships to take advantage of favorable conditions to proliferate, sometimes in unusual or extreme environments.

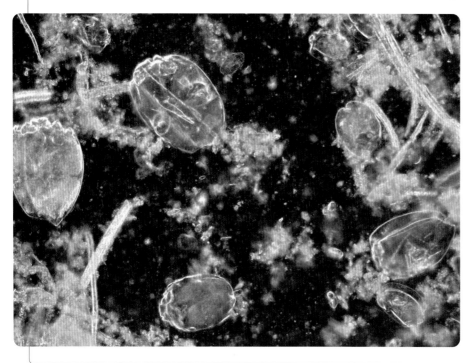

*Top* This bison calf is chewing the regurgitated contents of its stomach, called the cud.

*Left* Ciliates and bacteria live in the bison's stomach chamber, called the rumen. The microbes produce enzymes that digest the plant material, providing nutrients for the bison.

# Symbioses

Some microbes have coevolved to live together in a relationship called **symbiosis.** Sometimes one or both of the partners benefit from the association by obtaining nutrients, shelter, and/or protection from disease. When both partners benefit the symbiosis is called mutualistic. Other times a partner harms the host organism. This is a parasitic symbiosis.

## Microbes and Bison

Bison are **ruminants,** mammals that obtain nutrients from grass and other plants. Ruminants also include elk, deer, sheep, camels, giraffes, and cattle. Ruminants do not produce enzymes needed to digest **cellulose,** the tough material found in plant cell walls. Instead, they rely on symbiotic microbes to produce the enzymes to break down the otherwise unusable plant material and allow its conversion into nutritious food. Ruminants are completely dependent upon the microbes; without them, the ruminants would die.

Digestion occurs in the **rumen,** a large stomach chamber that contains billions of microbes. Ruminants chew a cud consisting of regurgitated, partially digested food mixed with saliva. The cud is churned in the rumen, where microbes—including anaerobic archaea, bacteria, fungi, and protozoa—ferment the plant matter into energy-rich compounds that pass into the gut, where they are absorbed. Carbon dioxide and methane are produced as by-products and are released into the atmosphere. Domestic livestock release about eighty million tons of methane annually, or about one-quarter of all global methane emissions. This potent greenhouse gas likely contributes to global climate changes.

## Fungi and Plants

**Endophytes** are specialized fungi that live inside plants. One grass found in Yellowstone's thermal areas has a fungus living within its roots and leaves. This grass, *Dichanthelium lanuginosum,* known as hot springs panic grass, is able to withstand soil

*Left* This rigid ciliate from a bison rumen is an entodiniomorph. At the front of the cell, two tufts of cilia are used for movement and feeding. A few fragments of plant material are inside the cell.

*Center* This highly magnified entodiniomorph cell has sharp spines.

*Right* A dense layer of cilia covers this type of ciliate called an isotrich, collected from a bison rumen.

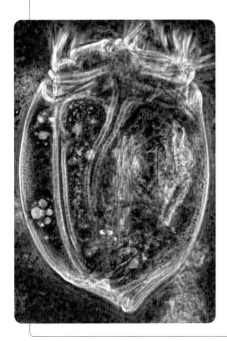
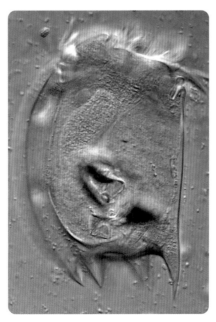
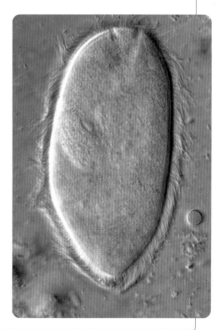

Left  *Dichanthelium lanuginosum* is a grass that normally grows in the tropics and southern areas of the United States, but it does quite well growing near hot springs and places with warm soil in Yellowstone National Park. *Image provided by R. Stout.*

Above  A fungus, called *Curvularia protuberata*, grows inside the roots of *Dichanthelium* plants. The fungus helps the plant survive in soils with temperatures that are too hot for most plants. *Image provided by K. Sheehan.*

Below left  Many plants form an association, called a mycorrhizae, with a fungus. The mycorrhizal fungi provide nutrients for the plant. In this image of a plant root, the white hyphae cover a brown root hair. *Image provided by R. Bunn and C. Zabinski.*

Below right  The fungus growing inside this root hair is stained with a blue dye so that it is visible under a microscope. *Image provided by R. Bunn and C. Zabinski.*

temperatures up to 55°C (131°F). This is much hotter than most grasses can tolerate. Without symbiotic fungus *Curvularia protuberata,* the grass cannot survive in high soil temperatures. Moreover, the fungus is unable to survive without its grass host. These two organisms have established a mutualistic relationship that allows both to grow in thermal areas in the park.

Many plants trade nutrients with fungi. Ectomycorrhizae are fungal-plant symbioses where the fungal partner of the symbiotic pair colonizes the outer layers of the plant roots. The ectomycorrhizal fungi are thought to provide phosphorous for the plant, while the plant provides nutrients to the fungus for growth. In poor soils, the fungal hyphae provide a greater surface area for the roots of the plants and aid in absorption of water and nutrients. Many of the ectomycorrhizal fungi not only colonize plant roots but also produce mushrooms on the soil surface.

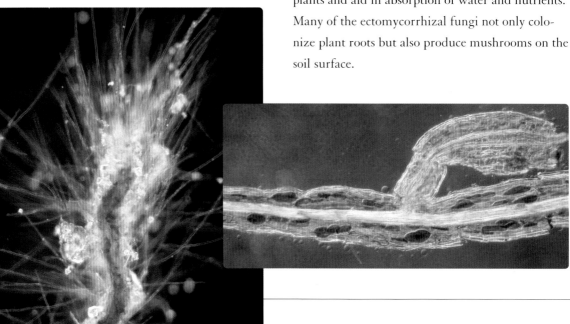

# Lichens

A lichen is another type of a mutualistic symbiosis between one or more photosynthetic microbes and a fungus. The photosynthetic partner may be either a green alga or a cyanobacterium or both. Many species of fungi have the ability to partner with algae to form lichens. The lichen looks very different from the individual algal, cyanobacterium, or fungal cells when they grow on their own. The algae and/or cyanobacteria carry out photosynthesis and produce nutrients for the fungal partner. The fungal partner helps to protect the algae from drying out or from high levels of ultraviolet light in the harsh habitats where lichens occur. The fungus also helps to acquire minerals that are used by the lichen for growth.

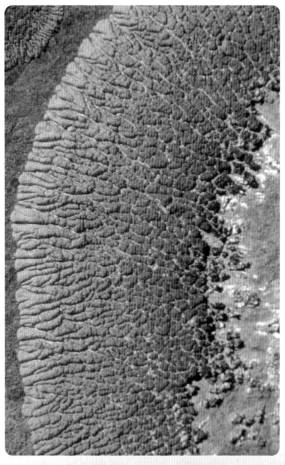

*Below* A lichen is a symbiotic association between a fungus and an alga. Different algae and fungi form unique lichens. Lichens are common on rocks, on tree bark, and in dry habitats throughout Yellowstone.

*Right* This brilliantly colored lichen grows on rocks in Yellowstone. *Image provided by S. Eversman.*

More than thirty thousand species of lichens on Earth grow in habitats with little moisture or nutrients. There are 186 described species of lichens in Yellowstone, where their brightly colored colonies are common on rocks, on trees, and in alpine regions. Most lichens grow slowly, 1 millimeter or less per year, depending upon the partners, the amount of moisture, the temperature, and the amount of sunlight they receive. A lichen 2 centimeters in diameter may be several years old.

## Algae and Protozoa

Algae form associations with many kinds of organisms, ranging from the corals of Australia's Great Barrier Reef to tree sloths. Many protozoa form symbiotic partnerships with algae. The ciliate *Paramecium bursaria* contains hundreds of round *Chlorella* cells. For a long time, scientists thought that this association benefited the *Paramecium* by providing nutrients from photosynthesis, a key to the *Paramecium*'s survival if no other food was available. More recently, it has been suggested that another, perhaps more important, benefit could be a supply of oxygen produced by the algae as a result of photosynthesis. This would allow the *Paramecium* to live in habitats that lack oxygen. In return the algae are protected and provided with nutrients inside the *Paramecium*.

## *Pilobolus*

*Pilobolus* is a fungus that lives in the intestinal tract of many of Yellowstone's animals, including moose, bighorn sheep, deer, pronghorn, bison, and elk.

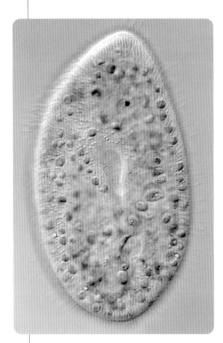

*Left* Paramecium bursaria is one of many ciliates that live symbiotically with green algae.

*Top* Elk, like bison, require microbes to digest the grass and other plants that they eat. Grazing animals often lie down while digestion takes place. *Image provided by D. Patterson.*

*Right* Elk droppings provide a nutritious food source for many microbes. *Image provided by D. Patterson.*

*Below right* Pilobolus is a fungus that colonizes the droppings of elk, deer, and bighorn sheep. These structures contain fungal spores in the black sporangium caps. *Image provided by M. Foos.*

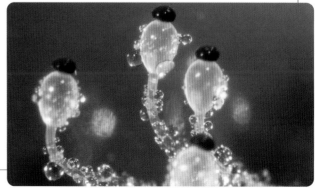

Spores of the fungus are excreted into animal dung, where they germinate and grow into fungal hyphae. While decomposing the dung, the fungal hyphae produce new spores in a black structure called a **sporangium.** Grazing animals ingest the spores and become reinfected with the fungus.

The larvae of a parasitic roundworm (nematode), *Dictyocaulus viviparus*, become parasites in the animals and infect their lungs. The larvae of the worms hatch in animal dung, are attracted to the water in the fungal sporangium, and crawl into the sporangium where they are discharged along with the fungal spores onto surrounding grass. Grazing animals then eat the grass that is covered with the larval parasites. The nematodes exploit the fungi to initiate a new round of lungworm infection.

# Microbial Mats

Microbial mats are prevalent in many thermal areas in Yellowstone. The mats contain communities of photosynthetic organisms or other pigmented microbes. Cyanobacteria and *Chloroflexus* (green nonsulfur bacterium) communities often grow in extensive mats that are visible to the naked eye. The mats develop slowly in and around hot, alkaline water habitats with temperatures about 70°C (158°F).

Mat-forming organisms are found in many habitats around the world, but large and extensive mats tend to occur only in habitats referred to as "extreme." In nonextreme environments, there are abundant protozoa and animal communities that eat the microbes and prevent the growth of large mats. The numerous extreme hot springs and their outflow channels in Yellowstone are favorable for the formation of mats, because the springs and channels are too hot for most other organisms. At another extreme location, Shark Bay in Western Australia, massive stromatolites, or columns of microbial mats that look like footstools, form in very salty water, a different kind of extreme environment.

Microbes form biofilms on almost any wet surface: rocks, aquatic environments, contact lenses, or even teeth (as dental plaque). The mats found around the thermal features in Yellowstone are biofilms on a considerably larger scale. Microbes secrete sticky sugar and protein polymers that hold biofilms together. The gelatinous material provides some protection for less tolerant species to survive under extreme conditions.

There are many types of microbial mats, from loosely formed clumps of algae to massive struc-

Extraordinary colors found at Black Sand Basin are created by microbial communities, black particles of obsidian, and sunlight reflected from the hot springs.

tures extending over large areas. Within the park, it is possible to see microbial mats that are green, yellow, red, brown, and other colors. They float in water or form on sediments, sinter, travertine, and rocks. Microbial mats can have a jellylike texture with colored layers and can vary in texture and color. The upper surfaces of a mat receive more sunlight and are made of brightly pigmented, photosynthetic microbes. Closer scrutiny, whether by a microscope or by examining genetic material, reveals that mats contain many species. The orange mats in Yellowstone are dominated by various species of filamentous bacteria and other rod-shaped cyanobacteria. However, even these mats include many other less abundant microbes.

Many mats develop slowly, beginning as thin biofilms that are little more than a single layer of cells thick. As they grow and as more sticky sugars and polymers are produced, the mats become sizable.

At times during the daylight hours, photosynthesis produces large quantities of oxygen bubbles in a mat. If the mat is submerged, the bubbles

*Above* In some places like this travertine cliff at Mammoth Hot Springs, microbial mats are massive.

*Below* At Cliff Geyser, the intense colors created by microbes and minerals create a bizarre landscape.

often break free and rise to the surface. Sometimes, however, the bubbles are caught within the slimy mat, and promote the formation of living stromatolites.

Not all mats are composed primarily of bacteria. Communities of eukaryotic microbes dominate other mats at temperatures between 30°C to 60°C (86°F to 140°F). The green alga *Zygogonium* forms dark red mats in areas such as Nymph Lake. *Cyanidium* mats occur in acidic features in Yellowstone.

Mats can be easily damaged, and because some develop very slowly, it may take a long time for new growth. Extensive microbial mats are rare on Earth. Cyanobacterial mats form in salty environments found along coastlines. Other mats are found in thermal areas like the hot springs and their outflow channels in Yellowstone. The park is one of the few places on Earth that protect these unique communities of organisms.

*Above* Microbial mats coat the bottom of a slow-moving outflow channel at La Duke Spring. Some mats float on top of the water.

*Below left* This slice through a mat collected at La Duke Spring looks like "microbial lasagna." The layers are formed by bacteria and trapped sand particles. The layers have different colors depending upon the bacterial communities that are present.

*Bottom left* This piece of a brown mat is composed mostly of cyanobacteria.

*Below right* This highly magnified scanning electron microscope image of a tiny piece of a microbial mat has a spongy texture created by intertwined bacteria. *Image provided by W. Inskeep and color enhanced by D. Patterson.*

*Bottom right* This flattened piece of a mat is mostly filamentous bacteria.

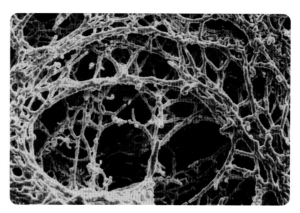

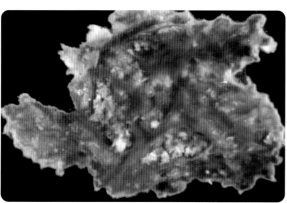

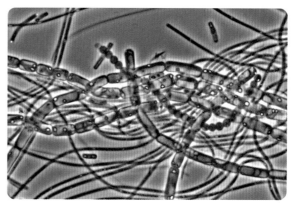

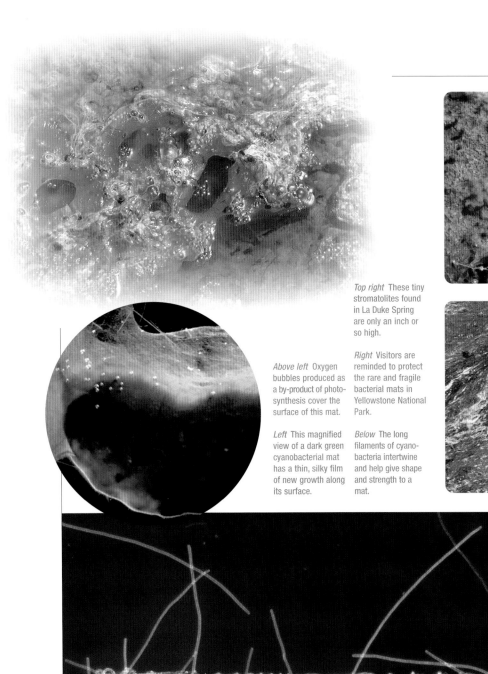
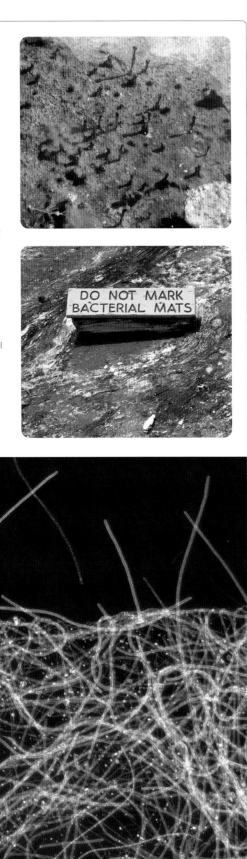

*Top right* These tiny stromatolites found in La Duke Spring are only an inch or so high.

*Above left* Oxygen bubbles produced as a by-product of photosynthesis cover the surface of this mat.

*Right* Visitors are reminded to protect the rare and fragile bacterial mats in Yellowstone National Park.

*Left* This magnified view of a dark green cyanobacterial mat has a thin, silky film of new growth along its surface.

*Below* The long filaments of cyanobacteria intertwine and help give shape and strength to a mat.

# Gradients

Microbes live in a world where environmental changes may take place dramatically and on a very small scale. Conditions that can affect the ability of microbes to survive and grow include temperature, nutrient availability, pH, presence or absence of oxygen, sunlight, moisture, and competition from other organisms and viruses. Variations in environmental conditions over time and/or distance are called gradients. Changes in gradients that occur within a distance of just a few millimeters can mean life or death for a microbe adapted to a specific niche if it is unable to respond to those

*Above* The band gradients of color were produced by communities of microbes that thrive in hot water flowing into Yellowstone Lake at West Thumb Geyser Basin.

*Below left* Extensive white sinter deposits frame Sunset Lake in Black Sand Basin. Photosynthetic bacteria create bands of color.

*Below right* A close-up shows distinct color zones created by mineral deposits and microbes that survive in this hot spring.

*The Relationships*

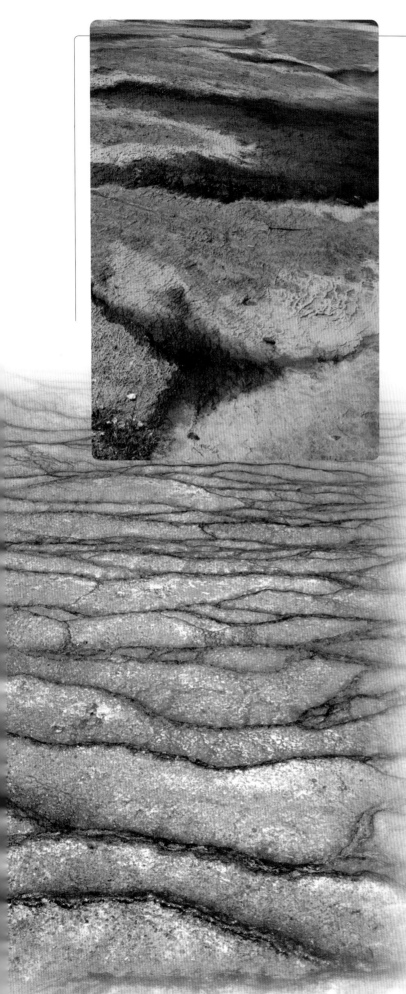

changes. Microbes may respond by becoming dormant, by producing resistant forms, or by moving to a new environment.

Sunlight greatly affects microbial growth in a mat community, providing the source of energy for photosynthetic organisms and releasing oxygen for use by other organisms. Photosynthetic microbes thrive in the upper parts of a mat where sunlight is able to penetrate. In the lower parts of a mat, where no light penetrates and oxygen levels are low, anaerobic bacteria may thrive. Light gradients can change quickly. During the night when there is no sunlight, photosynthesis stops and oxygen levels drop dramatically. The organisms within the mat must be tolerant of such changes if they are to survive. Some microbes change their metabolisms to take advantage of different conditions.

The most obvious gradients in Yellowstone are thermal gradients. A hot spring is often hottest in the center where subsurface water rises to the surface. As the water flows away from the source, it cools or degasses, and various chemical compounds dissolved in the hot water become insoluble and precipitate. As a result, not only does the temperature of the water change, but so does its chemistry. These changes create gradients around pools and their outflow channels. Different communities of microbes thrive in specific niches. Consequently, the changing conditions influence the composition and appearance of the microbial communities. Microbes often contain distinct pigments, creating bands of color in and around the pools and their outflow channels.

*Top left* Striking patterns of colors are created by microbial communities in this shallow area of a hot spring.

*Left* The combination of microbial populations and mineral deposits often create intricate, beautiful designs at Grand Prismatic Spring.

# Biomineralization

The formation of mineral deposits is dependent upon the chemistry of the subsurface water and the composition of subsurface rocks. Hot water that makes its way above ground is rich with compounds dissolved from these underlying rocks. The compounds often contain silicates, sulfur, iron, calcium carbonate, chlorides, and others. When the water reaches the surface, it degasses and evaporates. The dissolved compounds precipitate as colored deposits or crystals. Sometimes, in places like the terraces of Mammoth Hot Springs, this process is rapid, and visible formations appear within

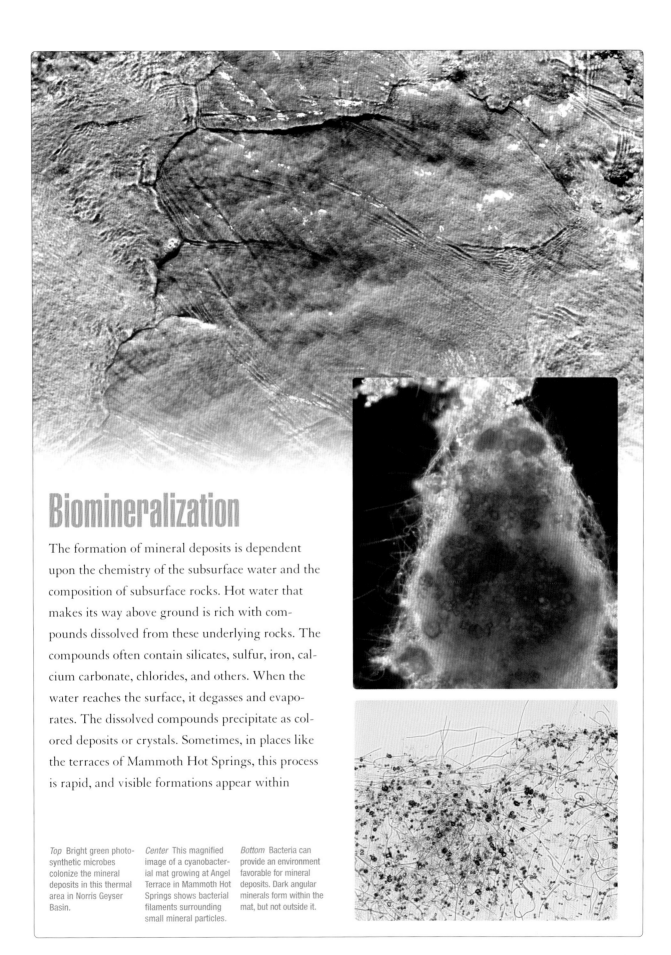

*Top* Bright green photosynthetic microbes colonize the mineral deposits in this thermal area in Norris Geyser Basin.

*Center* This magnified image of a cyanobacterial mat growing at Angel Terrace in Mammoth Hot Springs shows bacterial filaments surrounding small mineral particles.

*Bottom* Bacteria can provide an environment favorable for mineral deposits. Dark angular minerals form within the mat, but not outside it.

*The Relationships*

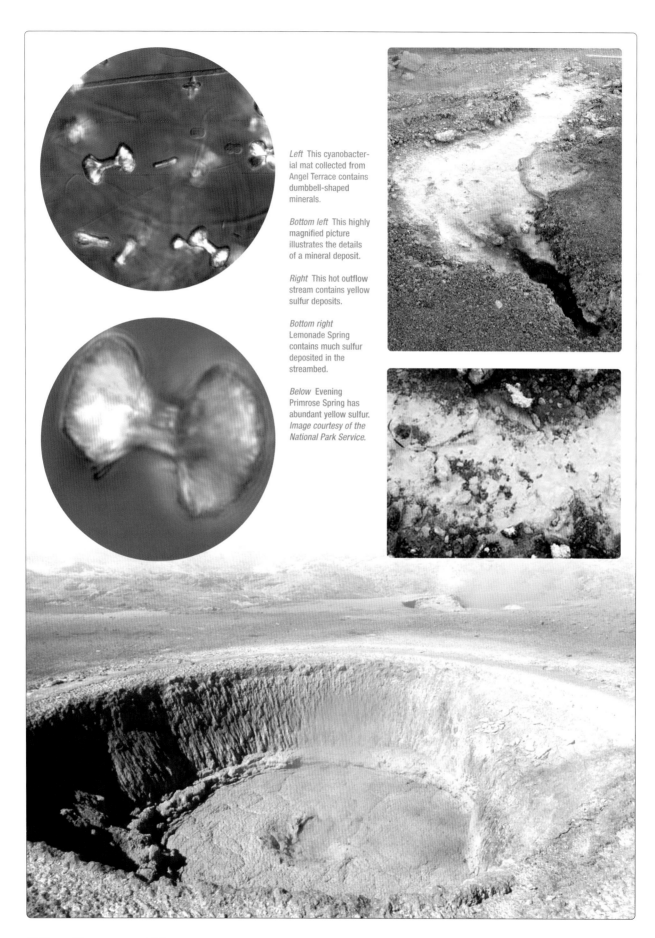

*Left* This cyanobacterial mat collected from Angel Terrace contains dumbbell-shaped minerals.

*Bottom left* This highly magnified picture illustrates the details of a mineral deposit.

*Right* This hot outflow stream contains yellow sulfur deposits.

*Bottom right* Lemonade Spring contains much sulfur deposited in the streambed.

*Below* Evening Primrose Spring has abundant yellow sulfur. *Image courtesy of the National Park Service.*

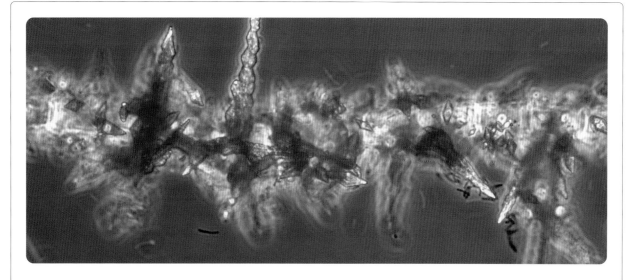

*Above* Sulfur crystals encrust this *Zygogonium* filament. The alga provides a surface for crystallization to begin. Dark rod-shaped bacteria are also present. *Below* These are magnified sulfur crystals.

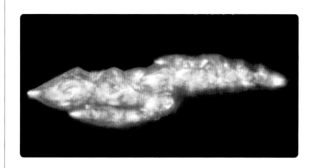

hours. In other features, such as Castle Geyser on Geyser Hill, the sinter cones develop over many hundreds of years.

Microbes can contribute to the formation of mineral deposits in a process called **biomineralization.** Often, as with the sulfur deposits in Lemonade Creek, there is a correlation between where the sediments form and the location of the microbial mats.

Microbial communities, for example, provide physical surfaces that aid in crystal formation. Microbes can change the chemical characteristics of an environment and hasten the precipitation of inorganic compounds or they may actively increase the concentration of certain compounds, and in so doing promote the formation of mineral deposits.

# Science and Yellowstone's Microbes

Yellowstone National Park attracts the attention of people from around the world. The appeal of geysers, mountain scenery, bison, birds, and flowers is obvious. Why is there so much interest by scientists in the microbes found in Yellowstone?

One of the most tantalizing challenges for the scientific community is to find out if life exists anywhere else in the universe. Scientists estimate that there are many other worlds that have temperature regimes and atmospheres that might be favorable for life. Many conclude that life must be abundant throughout the universe, even though firm evidence for this is elusive. Some biologists often consider that the evolutionary events that led to the emergence and evolution of life on Earth are so improbable that even if favorable conditions do occur in other places in the universe, the existence of life is highly unlikely.

The NASA astrobiology program studies the evolution of life on Earth, in the context of the history of the universe, to determine if life exists anywhere else. Much of the research is focused on the microbial world and on life in extreme environments like ones found in the hot springs of Yellowstone. One reason for studying extreme habitats is that microbes arose and diversified early in the Earth's history when there were high temperatures, no oxygen, and intense volcanic activity. These extreme conditions dominated the first two-thirds of the history of the planet. Similar conditions exist on other planets in the solar system. If life does exist elsewhere, then knowledge about the microbial world in extreme habitats on Earth may aid in investigations of life in remote worlds such as Mars or Europa, a moon of Jupiter.

Other scientists, often funded by the National Science Foundation or by NASA, are interested in the diversity of species and their physiological roles in the environment. Montana State University's Thermal Biology Institute focuses its research on the basic biology, geology, and ecology of hot springs in Yellowstone with a long-term goal of understanding how organisms respond and adapt to the unique physical and chemical features found in thermal environments.

Yellowstone National Park preserves one of the greatest concentrations of thermophilic biological diversity in the world. Thermophiles and other microbes have potential to benefit society by contributing to biotechnology, medicine, and basic scientific understanding. Enzymes isolated from microbes found in Yellowstone currently are used commercially in the production of ethanol, for bioremediation of contaminated wastes, for improving animal feed, in processes for recovering oil, for improving cleaning efficiencies of detergents, and in biological and medical laboratories. Thermophiles are considered by some to be the most important resource protected in Yellowstone.

"We saw beautiful things, beautiful enough for a poet's dream. I remember that we paused long by an inactive geyser whose pool was the perfection of geyser water—a great pure sparkling sapphire rippling with heat and catching sunlight on its undulations to reflect in broken gleams into its wondrous depths. It was bordered as far down as we could see by the softest of coral-like formations, white and pearl gray."

—MARGARET CRUIKSHANK, WRITING IN 1883

# Glossary

**Archaea:** One of the three main groups of organisms, sometimes referred to as Archaebacteria. Archaea are genetically and chemically distinct from the two other groups, the Eukarya and Eubacteria. Archaea cells lack nuclei and have unique membranes.

**Bacteria:** One of the three main groups of organisms, sometimes referred to as Eubacteria. Bacteria lack nuclei and are genetically and chemically distinct from the Eukarya and Archaea, the other two groups.

**Bacteriophage:** A virus that attacks bacteria.

**Biofilm:** A sticky coating formed by a single bacterial species or, more often, many species of bacteria, as well as fungi, algae, protozoa, debris, and corrosion products. Biofilms can form on any moist surface.

**Biomass:** The total weight of material produced by living organisms.

**Biomineralization:** A process in which living organisms form mineral crystals.

**Caldera:** A large, usually circular crater created when liquid rock erupts from a volcano.

**Carotenoid:** A colored chemical compound found in cells, sometimes used for protection from intense sunlight, but also used to capture sunlight energy during photosynthesis.

**Cellulose:** A material made by plants (and some bacteria) from glucose (sugar) molecules repeatedly joined together into chains (polymers). Cellulose is the major component in plant cell walls.

**Chlorophyll:** An energy-capturing substance used in the process of converting sunlight to energy.

**Chloroplast:** A compartment inside plant and algal cells where photosynthesis takes place.

**Chytrid:** A type of single-celled microbe (protist) that attacks other organisms and is related to the ancestors of the fungi.

**Cilia:** Hairlike projections from cells used either to move the cell itself or to move substances over or around the cell. A single one of these cell projections is called a *cilium*.

**Ciliate:** A type of protozoa; most have cilia all over the body.

**Cirrius:** Cilia used by ciliates like a leg to push against solid surfaces to help the organism move.

**Cocci:** Spherical structures; ball-shaped bacteria.

**Cyanobacteria:** A large group of photosynthetic and aquatic bacteria, often referred to as blue-green algae.

**Cyst:** A stage in the life of some organisms when the cell encloses itself within a protective wall and becomes inactive. Cysts often form when the environment becomes challenging, when water dries up, or when there is no food.

**Cytostome:** The mouth of a protozoan cell, the region where food is drawn into the cell and enclosed within a compartment known as a vacuole, where the food is digested.

**Desmid:** A green alga, typically appearing as if two cells are joined side by side.

**Detritus:** Dead and decaying debris found on riverbeds and in ponds.

**Diatom:** A type of single-celled alga whose body is surrounded with a wall of silica (glass).

**Domain:** A term used to describe one of the major branches of the tree of life. There are three domains: the Bacteria (Eubacteria), Archaea (Archaebacteria), and Eukarya (Eukaryota).

**Endophyte:** An organism that lives inside plants.

**Endospore:** A dormant structure inside bacterial cells that is highly resistant to heat, dryness, radiation, and disinfectants.

**Eukarya:** One of the three main branches of the tree of life. Eukarya are distinguished from both Archaea and Eubacteria because they have membrane-bound compartments (such as nuclei, chloroplasts, and mitochondria) inside the cell. Plants and animals belong to the Eukarya.

**Eukaryote:** The common term for a member of the Eukarya.

**Extrusome:** A compartment found inside eukaryotic cells that can be discharged to the outside of the cell and used for defense or to capture food.

**Filamentous:** Threadlike or hairlike.

**Flagella:** Long, hairlike structures on cells that help move cells through liquid or help move liquid past cells. A single one of these structures is called a *flagellum*.

**Flagellate:** A type of protozoa that moves or feeds using flagella.

**Frustule:** The glass shell of a diatom.

**Fumarole:** A vent or crack in the ground from which volcanic gases or steam escape into the atmosphere.

**Gradient:** A change in an environmental condition (such as temperature or light) with distance from the source. Temperature gradients vary widely over the earth, sometimes increasing dramatically around volcanic areas.

**Habitat:** The place or environment where an organism lives.

**Halophile:** An organism that lives in a very salty environment.

**Hyperthermophile:** An organism that lives in environments above 80°C (176°F).

**Hyphae:** The long, threadlike strand of a fungus that feeds, grows, and may produce a mushroom or some other kind of reproductive structure. A single strand is too small to be seen with the naked eye and is called a *hypha*.

**Magma plume:** A column of liquid rock rising from Earth's interior.

**Mat:** A thick, layered coating (biofilm) formed by communities of microbes.

**Metabolism:** The chemical processes within living cells.

**Methanogen:** A microbe that lives without oxygen and releases methane gas as a waste product of cellular metabolism.

**Microbe:** A microscopic organism such as a bacterium (Eubacteria and Archaea), virus, alga, fungus, and protist.

**Micrograph:** A photograph taken with a camera attached to a microscope.

**Mineralization:** A geological process that produces hard crystalline materials such as limestone, sulfur, or salt or soft materials like clay. Living organisms make mineral compounds in a process called biomineralization. Microbes secrete substances such as sulfur or iron that create surfaces favorable for crystal formation. Fossil stromatolites, teeth, bones, kidney stones, seashells, and coral also are examples of biominerals.

**Mitochondria:** Compartments inside eukaryotic cells where much of the energy for metabolism is produced. A single one of these compartments is called a *mitochondrion*.

**Mutation:** A genetic change or alteration that can be passed on to offspring.

**Mycelium:** A network or web of strands (hyphae) produced by fungi.

**Nitrates:** Nitrogen-based chemicals required for making proteins.

**Organelle:** A compartment within a eukaryotic cell, surrounded by one or two membranes, where special functions such as photosynthesis are carried out.

**Pennate:** A type of long, diamond-shaped diatom (single-celled alga).

**Photosynthesis:** A process by which organisms capture sunlight and transfer the energy into chemical compounds for use by the cell.

**Phycocyanin:** A photosynthetic pigment, commonly found in cyanobacteria.

**Pigment:** A colored chemical compound.

**Polymerase chain reaction (PCR):** A laboratory procedure for making copies of DNA or RNA, the genetic material present in all living organisms.

**Precipitate:** To separate from a liquid as a solid.

**Prokaryote:** Single-celled microorganisms without a membrane-bound nucleus; the Archaea and Bacteria.

**Protist:** A single-celled eukaryotic microbe such as an alga or amoeba.

**Pseudopodia:** Temporary, footlike extensions from an amoeba, used for movement or to capture food; sometimes called "false feet." A single one of these feet is a *pseudopodium*.

**Raphe:** A slit or groove along the top and bottom sides of pennate diatoms (diamond-shaped, single-celled algae). Raphes are associated with cell movement.

**Rhizoid:** Fine threadlike extensions from some kinds of cells, used to capture food and nutrients.

**Ribosomal RNA:** RNA (ribonucleic acid) is a chemical found in the nucleus and cytoplasm of cells that plays an important role in protein synthesis and other chemical activities of the cell. The structure of RNA is similar to that of DNA. There are several classes of RNA molecules, including messenger RNA, transfer RNA, ribosomal RNA (found in cell structures called ribosomes), and other small RNAs, each serving a different purpose.

**Rumen:** A special type of stomach chamber where millions of microbes digest the plant material (cellulose) in grass and shrubs, providing nutrients for grazing mammals.

**Ruminants:** Cud-chewing mammals such as cattle, sheep, goats, deer, elk, and bison that have a special stomach called a rumen.

**Saprophyte:** An organism, especially a fungus or bacterium, that eats dead organic material.

**Sinter:** A mineral deposit from hot springs and geysers, occurring as an incrustation around the springs, and sometimes forming conical mounds or terraces.

**Solfatara:** A volcanic area or steam vent with high levels of sulfur.

**Sporangium:** A structure in fungi that is used to hold and distribute spores. Multiples of these structures are called *sporangia*.

**Streamer:** Wispy, hairlike strands formed by colonies of bacteria and archaea.

**Stromatolite:** A layered, vertical column or domed structure formed by cyanobacteria.

**Symbiont:** An organism that lives with another organism.

**Symbiosis:** The interaction of two or more organisms.

**Thermal feature:** A geyser, fumarole, or spring heated by underlying molten rock in a volcanic region.

**Thermoacidophile:** An organism that lives in hot, acidic environments.

**Thermophile:** An organism that lives in hot environments.

**Travertine:** A calcium carbonate mineral deposit.

**Trichocyst:** A kind of compartment (organelle) found in a type of single-celled microbe (a protist) that can be expelled from a cell as a stiff thread.

**Vacuole:** A compartment in a cell that is usually filled with liquid and is surrounded by a membrane.

**Virus:** An extremely small piece of nucleic acid (DNA, RNA, or membrane material) wrapped in a protective coat of protein that reproduces only within cells of living hosts.

# Resources

## Books

This book provides a glimpse into the world of microbes found in Yellowstone National Park, a world that harbors an immense diversity of life in some of the most remarkable habitats on Earth. If you wish to learn more about microbes, Yellowstone National Park, or other topics found in this book, we suggest the following:

*Biodiversity of Microbial Life: Foundations of Earth's Biosphere,* edited by James T. Staley and Anna-Louise Reysenbach. New York: Wiley-Liss, Inc., 2002.

*Brock Biology of Microorganisms*, tenth edition, by M. T. Madigan, J. M. Martinko, and J. Parker. Englewood Cliffs, N.J.: Prentice Hall, 2003.

*Free-living Freshwater Protozoa: A Color Guide,* by David J. Patterson. Washington, D.C.: ASM Press, 2003.

*The Geysers of Yellowstone,* third edition, by T. Scott Bryan. Boulder: University Press of Colorado, 2003.

*Das Leben in Wassertropfen,* by H. Streble and D. Krauter. Stuttgart: Franch'sche Verlag, 1976.

*Lichens,* by William Purvis. Washington, D.C.: The Smithsonian Institution Press, 2000.

*Life at High Temperatures,* by Thomas D. Brock. Yellowstone National Park, Wyo.: Yellowstone Association for Natural Science, History and Education, Inc., 1994.

*Protozoa and Other Protists,* by M. A. Sleigh. London: Arnold, 1989.

*Windows into the Earth,* by Robert B. Smith and Lee J. Siegel. London: Oxford University Press, 2000.

## Useful Web Sites

micro∗scope: http://microscope.mbl.edu

The Thermal Biology Institute, Montana State University: http://tbi.montana.edu

Scientific research and science in Yellowstone Park: http://www.wsulibs.swu.edu/yellowstone

Yellowstone National Park, National Park Service: http://www.nps.gov/yell

Yellowstone Volcano Observatory: http://volcanoes.usgs.gov/yvo/

# Contributors

## Authors

Kathy B. Sheehan
Division of Health Sciences
WWAMI Medical Education Program
Montana State University
308 Leon Johnson Hall
Bozeman, MT 59717
(406) 994-5415
umbks@montana.edu
Images pages xi (right), 21, 88 (top right)

David J. Patterson
Bay Paul Center
Marine Biological Laboratory
Woods Hole, MA 02543
(508) 289-7260
dpatterson@mbl.edu
Microscopic images, except where noted, and images pages 7, 90 (top and right)

Brett Leigh Dicks
Biology Department
Santa Barbara City College
Santa Barbara, CA 93109
(805) 965-0581, ext. 2319
dicks@sbcc.edu
Landscape images, except where noted

Joan M. Henson
Department of Microbiology
Montana State University
109 Lewis Hall
Bozeman, MT 59717
(406) 994-4690
jhenson@montana.edu

## Additional Illustrations and Figures

### ● Montana State University

Mark Young
Department of Plant Sciences
Bozeman, MT 59717
(406) 994-5158
myoung@montana.edu
Images page 60 (bottom)

Sue Brumfield
Department of Plant Sciences
Bozeman, MT 59171
(406) 994-5144
uplsb@montana.edu
Image page 60 (bottom)

George Rice
Department of Plant Sciences
Bozeman, MT 59717
(406) 994-5146
grice@montana.edu
Image page 60 (top)

William Inskeep
Land Resources and
Environmental Sciences
Bozeman, MT 59717
(406) 994-5077
binskeep@montana.edu
Images page 68 (middle), 93 (top right)

Jessie Donohoe-Christiansen
Land Resources and Environmental Sciences
Bozeman, MT 59717
(406) 994–2190
Image page 61 (bottom right inset)

Sharon Eversman
Department of Ecology
Bozeman, MT 59717
(406) 994–2473
eversman@montana.edu
Image page 89 (top right)

Cathy Zabinski
Land Resources and Environmental Sciences
Bozeman, MT 59717
(406) 994–4227
cathyz@montana.edu
Images page 88 (bottom left and right)

Rebecca Bunn
Land Resources and Environmental Sciences
Bozeman, MT 59717
(406) 994–4227
Images page 88 (bottom left and right)

Rich Stout
Department of Plant Sciences
Bozeman, MT 59717
(406) 994–4912
rstout@montana.edu
Image page 88 (top left)

## Indiana Univerity East

K. Michael Foos
Department of Biology, Whitewater Hall, Room 273
Richmond, IN 47374
(765) 973–8303
foos@indiana.edu
Image page 90 (bottom right)

## Yellowstone National Park

Jim Peaco, Park Photographer
Albright Visitor Center
P.O. Box 168
Yellowstone National Park, WY 82190
(307) 344–2266
jim_peaco@nps.gov
Images pages 37 (top), 54 (bottom), 98 (bottom)

Adam Kiel and Shannon Savage
Spatial Analysis Center
P.O. Box 168
Yellowstone National Park, WY 82190
(307) 344–2246
Adam_Kiel@nps.gov
All maps

## Graphic Design/ Tree of Life

David Adams Thompson, Inc.
Jason King
1303 South Montana Avenue
Bozeman, MT 59715
(406) 586–3871
thompson@montana.com
Chart page 57